Lecture Notes in Computer Science 3000

Commenced Publication in 1973
Founding and Former Series Editors:
Gerhard Goos, Juris Hartmanis, and Jan van Leeuwen

Springer
Berlin
Heidelberg
New York
Hong Kong
London
Milan
Paris
Tokyo

Martin Dietzfelbinger

Primality Testing in Polynomial Time

From Randomized Algorithms to "PRIMES is in P"

Springer

Author

Martin Dietzfelbinger
Technische Universität Ilmenau
Fakultät für Informatik und Automatisierung
98684 Ilmenau, Germany
E-mail: martin.dietzfelbinger@tu-ilmenau.de

Library of Congress Control Number: 2004107785

CR Subject Classification (1998): F.2.1, F.2, F.1.3, E.3, G.3

ISSN 0302-9743
ISBN 3-540-40344-2 Springer-Verlag Berlin Heidelberg New York

Springer-Verlag is a part of Springer Science+Business Media

springeronline.com

© Springer-Verlag Berlin Heidelberg 2004
Printed in Germany

Typesetting: Camera-ready by author, data conversion by Boller Mediendesign
Printed on acid-free paper   SPIN: 10936009   06/3142   5 4 3 2 1 0

To Angelika, Lisa, Matthias, and Johanna

Preface

On August 6, 2002, a paper with the title "PRIMES is in P", by M. Agrawal, N. Kayal, and N. Saxena, appeared on the website of the Indian Institute of Technology at Kanpur, India. In this paper it was shown that the *"primality problem"* has a *"deterministic algorithm"* that runs in *"polynomial time"*.

Finding out whether a given number n is a prime or not is a problem that was formulated in ancient times, and has caught the interest of mathematicians again and again for centuries. Only in the 20th century, with the advent of cryptographic systems that actually used large prime numbers, did it turn out to be of practical importance to be able to distinguish prime numbers and composite numbers of significant size. Readily, algorithms were provided that solved the problem very efficiently and satisfactorily for all practical purposes, and provably enjoyed a time bound polynomial in the number of digits needed to write down the input number n. The only drawback of these algorithms is that they use *"randomization"* — that means the computer that carries out the algorithm performs random experiments, and there is a slight chance that the outcome might be wrong, or that the running time might not be polynomial. To find an algorithm that gets by without randomness, solves the problem error-free, and has polynomial running time had been an eminent open problem in complexity theory for decades when the paper by Agrawal, Kayal, and Saxena hit the web. The news of this amazing result spread very fast around the world among scientists interested in the theory of computation, cryptology, and number theory; within days it even reached The New York Times, which is quite unusual for a topic in theoretical computer science.

Practically, not much has changed. In cryptographic applications, the fast randomized algorithms for primality testing continue to be used, since they are superior in running time and the error can be kept so small that it is irrelevant for practical applications. The new algorithm does not seem to imply that we can factor numbers fast, and no cryptographic system has been broken. Still, the new algorithm is of great importance, both because of its long history and because of the methods used in the solution.

As is quite common in the field of number-theoretic algorithms, the formulation of the deterministic primality test is very compact and uses only very simple basic procedures. The analysis is a little more complex, but astoundingly it gets by with a small selection of the methods and facts taught in introductory algebra and number theory courses. On the one hand, this raises the philosophical question whether other important open problems in theoretical computer science may have solutions that require only basic methods. On the other hand, it opens the rare opportunity for readers without a specialized mathematical training to fully understand the proof of a new and important result.

It is the main purpose of this text to guide its reader all the way from the definitions of the basic concepts from number theory and algebra to a full understanding of the new algorithm and its correctness proof and time analysis, providing details for all the intermediate steps. Of course, the reader still has to go the whole way, which may be steep in some places; some basic mathematical training is required and certainly a good measure of perseverance.

To make a contrast, and to provide an introduction to some practically relevant primality tests for the complete novice to the field, also two of the classical primality testing algorithms are described and analyzed, viz., the "Miller-Rabin Test" and the "Solovay-Strassen Test". Also for these algorithms and their analysis, all necessary background is provided.

I hope that this text makes the area of primality testing and in particular the wonderful new result of Agrawal, Kayal, and Saxena a little easier to access for interested students of computer science, cryptology, or mathematics.

I wish to thank the students of two courses in complexity theory at the Technical University of Ilmenau, who struggled through preliminary versions of parts of the material presented here. Thanks are due to Juraj Hromkovič for proposing that this book be written as well as his permanent encouragement on the way. Thomas Hofmeister and Juraj Hromkovič read parts of the manuscript and gave many helpful hints for improvements. (Of course, the responsibility for any errors remains with the author.) The papers by D.G. Bernstein, generously made accessible on the web, helped me a lot in shaping an understanding of the subject matter. I wish to thank Alfred Hofmann of Springer-Verlag for his patience and the inexhaustible enthusiasm with which he accompanied this project. And, finally, credit is due to M. Agrawal, N. Kayal, and N. Saxena, who found this beautiful result.

Ilmenau, March 2004   *Martin Dietzfelbinger*

Contents

1. Introduction: Efficient Primality Testing

1.1 Algorithms for the Primality Problem

A natural number $n > 1$ is called a **prime number** if it has no positive divisors other than 1 and n. If n is not prime, it is called **composite**, and can be written $n = a \cdot b$ for natural numbers $1 < a, b < n$. Ever since this concept was defined in ancient Greece, the **primality problem**

"Given a number n, decide whether n is a prime number or not"

has been considered a natural and intriguing computational problem. Here is a simple algorithm for the primality problem:

Algorithm 1.1.1 (Trial Division)

INPUT: Integer $n \geq 2$.
METHOD:
```
0      i: integer;
1      i ← 2;
2      while i · i ≤ n repeat
3          if i divides n
4              then return 1;
5          i ← i + 1;
6      return 0;
```

This algorithm, when presented with an input number n, gives rise to the following calculation: In the loop in lines 2–5 the numbers $i = 2, 3, \ldots, \lfloor\sqrt{n}\rfloor$, in this order, are tested for being a divisor of n. As soon as a divisor is found, the calculation stops and returns the value 1. If no divisor is found, the answer 0 is returned. The algorithm solves the primality problem in the following sense:

n is a prime number   if and only if   Algorithm 1.1.1 returns 0.

This is because if $n = a \cdot b$ for $1 < a, b < n$, then one of the factors a and b is not larger than $\sqrt{n}$, and hence such a factor must be found by the algorithm. For moderately large n this procedure may be used for a calculation by hand; using a modern computer, it is feasible to carry it out for numbers with 20 or 25 decimal digits. However, when confronted with a number like

M. Dietzfelbinger: Primality Testing in Polynomial Time, LNCS 3000, pp. 1-12, 2004.

$n = 7483845764874895490005046457879234760435948750902645 2654305481,$

this method cannot be used, simply because it takes too long. The 62-digit number n happens to be prime, so the loop runs for more than 10^{31} rounds. One might think of some simple tricks to speed up the computation, like dividing by 2, 3, and 5 at the beginning, but afterwards not by any proper multiples of these numbers. Even after applying tricks of this kind, and under the assumption that a very fast computer is used that can carry out one trial division in 1 nanosecond, say, a simple estimate shows that this would take more than 10^{13} years of computing time on a single computer.

Presented with such a formidably large number, or an even larger one with several hundred decimal digits, naive procedures like trial division are not helpful, and will never be even if the speed of computers increases by several orders of magnitude and even if computer networks comprising hundreds of thousands of computers are employed.

One might ask whether considering prime numbers of some hundred decimal digits makes sense at all, because there cannot be any set of objects in the real world that would have a cardinality that large. Interestingly, in algorithmics and especially in cryptography there are applications that use prime numbers of that size for very practical purposes. A prominent example of such an application is the public key cryptosystem by Rivest, Shamir, and Adleman [36] (the "*RSA system*"), which is based on our ability to create random primes of several hundred decimal digits. (The interested reader may wish to consult cryptography textbooks like [37, 40] for this and other examples of cryptosystems that use randomly generated large prime numbers.)

One may also look at the primality problem from a more theoretical point of view. A long time before prime numbers became practically important as basic building blocks of cryptographic systems, Carl Friedrich Gauss had written:

> "*The problem of distinguishing prime numbers from composites, and of resolving composite numbers into their prime factors, is one of the most important and useful in all of arithmetic. ... The dignity of science seems to demand that every aid to the solution of such an elegant and celebrated problem be zealously cultivated.*" ([20], in the translation from Latin into English from [25])

Obviously, Gauss knew the trial division method and also methods for finding the prime decomposition of natural numbers. So it was not just any procedure for deciding primality he was asking for, but one with further properties — simplicity, maybe, and speed, certainly.

1.2 Polynomial and Superpolynomial Time Bounds

In modern language, we would probably say that Gauss asked for an *efficient* algorithm to test a number for being a prime, i.e., one that solves the problem *fast* on numbers that are *not too large*. But what does "*fast*" and "*not too large*" mean? Clearly, for any algorithm the number of computational steps made on input n will grow as larger and larger n are considered. It is the *rate* of growth that is of interest here.

To illustrate a growth rate different from $\sqrt{n}$ as in Algorithm 1.1.1, we consider another algorithm for the primality problem (Lehmann [26]).

Algorithm 1.2.1 (Lehmann's Primality Test)

INPUT: Odd integer $n \geq 3$, integer $\ell \geq 2$.
METHOD:

```
0      a, c: integer; b[1..ℓ]: array of integer;
1      for i from 1 to ℓ do
2          a ← a randomly chosen element of {1,...,n − 1};
3          c ← a^(n−1)/2 mod n;
4          if c ∉ {1, n − 1}
5              then return 1;
6              else b[i] ← c;
7      if b[1] = ··· = b[ℓ] = 1
8          then return 1;
9          else return 0;
```

The intended output of Algorithm 1.2.1 is 0 if n is a prime number and 1 if n is composite. The loop in lines 1–6 causes the same action to be carried out ℓ times, for $\ell \geq 2$ a number given as input. The core of the algorithm is lines 2–6. In line 2 a method is invoked that is important in many efficient algorithms: **randomization**. We assume that the computer that carries out the algorithm has access to a source of randomness and in this way can choose a number a in $\{1,\ldots,n-1\}$ uniformly at random. (Intuitively, we may imagine it casts fair "dice" with $n-1$ faces. In reality, of course, some mechanism for generating "pseudorandom numbers" is used.) In the ith round through the loop, the algorithm chooses a number a_i at random and calculates $c_i = a_i^{(n-1)/2} \bmod n$, i.e., the remainder when $a_i^{(n-1)/2}$ is divided by n. If c_i is different from 1 and $n-1$, then output 1 is given, and the algorithm stops (lines 4 and 5); otherwise (line 6) c_i is stored in memory cell $b[i]$. If all of the c_i's are in $\{1, n-1\}$, the loop runs to the end, and in lines 7–9 the outcomes $c_1,\ldots,c_\ell$ of the ℓ rounds are looked at again. If $n-1$ appears at least once, output 0 is given; if all c_i's equal 1, output 1 is given.

We briefly discuss how the output should be interpreted. Since the algorithm performs random experiments, the result is a random variable. What is the probability that we get the "wrong" output? We must consider two cases.

Case 1: n is a prime number. (The desired output is 0.) — We shall see later (Sect. 6.1) that for n an odd prime exactly half of the elements a of $\{1,\ldots,n-1\}$ satisfy $a^{(n-1)/2} \bmod n = n-1$, the other half satisfies $a^{(n-1)/2} \bmod n = 1$. This means that the loop runs through all ℓ rounds, and that the probability that $c_1 = \cdots = c_\ell = 1$ and the wrong output 1 is produced is $2^{-\ell}$.

Case 2: n is a composite number. (The desired output is 1.) — There are two possibilities. If there is no a in $\{1, \ldots, n-1\}$ with $a^{(n-1)/2} \bmod n = n-1$ at all, the output is guaranteed to be 1, which is the "correct" value. On the other hand, it can be shown (see Lemma 5.3.1) that if there is some a in $\{1, \ldots, n-1\}$ that satisfies $a^{(n-1)/2} \bmod n = n-1$, then more than half of the elements in $\{1, \ldots, n-1\}$ satisfy $a^{(n-1)/2} \bmod n \notin \{1, n-1\}$. This means that the probability that the loop in lines 1–6 runs for ℓ rounds is no more than $2^{-\ell}$. The probability that output 0 is produced cannot be larger than this bound.

Overall, the probability that the wrong output appears is bounded by $2^{-\ell}$. This can be made very small at the cost of a moderate number of repetitions of the loop.

Algorithm 1.2.1, our first "efficient" primality test, exhibits some features we will find again and again in such algorithms: the algorithm itself is very simple, but its correctness or error probability analysis is based on facts from number theory referring to algebraic structures not appearing in the text of the algorithm.

Now let us turn to the computational effort needed to carry out Algorithm 1.2.1 on an input number n. Obviously, the only interesting part of the computation is the evaluation of $a^{(n-1)/2} \bmod n$ in line 3. By "modular arithmetic" (see Sect. 3.3) we can calculate with remainders modulo n throughout, which means that only numbers of size up to n^2 appear as intermediate results. Calculating $a^{(n-1)/2}$ in the naive way by $(n-1)/2 - 1$ multiplications is hopelessly inefficient, even worse than the naive trial division method. But there is a simple trick ("repeated squaring", explained in detail in Sect. 2.3) which leads to a method that requires at most $2\log n$ multiplications[1] and divisions of numbers not larger than n^2. How long will this take in terms of single-digit operations if we calculate using decimal notation for integers? Multiplying an h-digit and an ℓ-digit number, by the simplest methods as taught in school, requires not more than $h \cdot \ell$ multiplications and $c_0 \cdot h \cdot \ell$ additions of single decimal digits, for some small constant c_0. The number $\|n\|_{10}$ of decimal digits of n equals $\lceil \log_{10}(n+1) \rceil \approx \log_{10} n$, and thus the number of elementary operations on digits needed to carry out Algorithm 1.2.1 on an n-digit number can be estimated from above by $c(\log_{10} n)^3$ for a suitable constant c. We thus see that Algorithm 1.2.1 can be carried out on a fast computer in reasonable time for numbers with several thousand digits.

As a natural measure of the *size of the input* we could take the number $\|n\|_{10} = \lceil \log_{10}(n+1) \rceil \approx \log_{10} n$ of decimal digits needed to write down n. However, closer to the standard representation of natural numbers in computers, we take the number $\|n\| = \|n\|_2 = \lceil \log(n+1) \rceil$ of digits of the binary rep-

---

[1] In this text, $\ln x$ denotes the logarithm of x to the base e, while $\log x$ denotes the logarithm of x to the base 2.

resentation of n, which differs from $\log n$ by at most 1. Since $(\log n)/(\log_{10} n)$ is the constant $(\ln 10)/(\ln 2) \approx 3.322 \approx \frac{10}{3}$, we have $\|n\|_{10} \approx \frac{3}{10} \log n$. For example, a number with 80 binary digits has about 24 decimal digits.) Similarly, as an elementary operation we view the addition or the multiplication of two bits. A rough estimate on the basis of the naive methods shows that certainly $c \cdot \ell$ bit operations are sufficient to add, subtract, or compare two ℓ-bit numbers; for multiplication and division we are on the safe side if we assume an upper bound of $c \cdot \ell^2$ bit operations, for some constant c. Assume now an algorithm $\mathcal{A}$ is given that performs $T_{\mathcal{A}}(n)$ elementary operations on input n. We consider possible bounds on $T_{\mathcal{A}}(n)$ expressed as $f_i(\log n)$, for some functions $f_i : \mathbb{N} \to \mathbb{R}$; see Table 1.1. The table lists the bounds we get for numbers with about 60, 150, and 450 decimal digits, and it gives the binary length of numbers we can treat within 10^{12} and 10^{20} computational steps.

i	$f_i(x)$	$f_i(200)$	$f_i(500)$	$f_i(1500)$	$s_i(10^{12})$	$s_i(10^{20})$
1	$c \cdot x$	$200c$	$500c$	$1{,}500c$	$10^{12}/c$	$10^{20}/c$
2	$c \cdot x^2$	$40{,}000c$	$250{,}000c$	$2.2c \cdot 10^6$	$10^6/\sqrt{c}$	$10^{10}/\sqrt{c}$
3	$c \cdot x^3$	$8c \cdot 10^6$	$1.25c \cdot 10^8$	$3.4c \cdot 10^9$	$10^4/\sqrt[3]{c}$	$4.6 \cdot 10^6/\sqrt[3]{c}$
4	$c \cdot x^4$	$1.6c \cdot 10^9$	$6.2c \cdot 10^{10}$	$5.1c \cdot 10^{12}$	$1{,}000/\sqrt[4]{c}$	$100{,}000/\sqrt[4]{c}$
5	$c \cdot x^6$	$6.4c \cdot 10^{13}$	$1.6c \cdot 10^{16}$	$1.1c \cdot 10^{19}$	$100/\sqrt[6]{c}$	$2150/\sqrt[6]{c}$
6	$c \cdot x^9$	$5.1c \cdot 10^{20}$	$2.0c \cdot 10^{24}$	$3.8c \cdot 10^{28}$	$22/\sqrt[9]{c}$	$165/\sqrt[9]{c}$
7	$x^{2\ln\ln x}$	$4.7 \cdot 10^7$	$7.3 \cdot 10^9$	$4.3 \cdot 10^{12}$	$1{,}170$	$22{,}000$
8	$c \cdot 2^{\sqrt{x}}$	$18{,}000c$	$5.4c \cdot 10^6$	$4.55c \cdot 10^{11}$	$1{,}600$	$4{,}400$
9	$c \cdot 2^{x/2}$	$1.3c \cdot 10^{30}$	$1.6c \cdot 10^{60}$	$2.6c \cdot 10^{120}$	80	132

Table 1.1. Growth functions for operation bounds. $f_i(200)$, $f_i(500)$, $f_i(1500)$ denote the bounds obtained for 200-, 500-, and 1500-bit numbers; $s_i(10^{12})$ and $s_i(10^{20})$ are the maximal numbers of binary digits admissible so that an operation bound of 10^{12} resp. 10^{20} is guaranteed

We may interpret the figures in this table in a variety of ways. Let us (very optimistically) assume that we run our algorithm on a computer or a computer network that carries out 1,000 bit operations in a nanosecond. Then 10^{12} steps take about 1 second (feasible), and 10^{20} steps take a little more than 3 years (usually unfeasible). Considering the rows for f_1 and f_2 we note that algorithms that take only a linear or quadratic number of operations can be run for extremely large numbers within a reasonable time. If the bounds are cubic (as for Algorithm 1.2.1, f_3), numbers with thousands of digits pose no particular problem; for polynomials of degree 4 (f_4), we begin to see a limit: numbers with 30,000 decimal digits are definitely out of reach. Polynomial operation bounds with larger exponents (f_5 or f_6) lead to situations where the length of the numbers that can be treated is already severely restricted — with $(\log n)^9$ operations we may deal with one 7-digit number in 1 second; treating a single 50-digit number takes years. Bounds $f_7(\log n) = (\log n)^{2 \ln \ln n}$, $f_8(\log n) = c \cdot 2^{\sqrt{\log n}}$, and $f_9(\log n) = c\sqrt{n}$ exceed any polynomial in $\log n$ for sufficiently large n. For numbers with small binary length $\log n$, however, some of these superpolynomial bounds may still be smaller than high-degree polynomial bounds, as the comparison between f_6, f_7, and f_8 shows. In particular, note that for $\log n = 180,000$ (corresponding to a 60,000-digit number) we have $2 \ln \ln(\log n) < 5$, so $f_7(\log n) < (\log n)^5$.

The bound $f_9(\log n) = c\sqrt{n}$, which belongs to the trial division method, is extremely bad; only very short inputs can be treated.

Summing up, we see that algorithms with a polynomial bound with a truly small exponent are useful even for larger numbers. Algorithms with polynomial time bounds with larger exponents may become impossible to carry out even for moderately large numbers. If the time bound is superpolynomial, treating really large inputs is usually out of the question. From a theoretical perspective, it has turned out to be useful to draw a line between computational problems that admit algorithms with a polynomial operation bound and problems that do not have such algorithms, since for large enough n, every polynomial bound will be smaller than every superpolynomial bound. This is why the class P, to be discussed next, is of such prominent importance in computational complexity theory.

1.3 Is PRIMES in P?

In order to formulate what exactly the question "Is PRIMES in P?" means, we must sketch some concepts from computational complexity theory. Traditionally, the objects of study of complexity theory are "languages" and "functions". A nonempty finite set Σ is regarded as an **alphabet**, and one considers the set Σ^* of all finite sequences or **words** over Σ. The most important alphabet is the binary alphabet $\{0, 1\}$, where Σ^* comprises the words

$$\varepsilon \text{ (the empty word)}, 0, 1, 00, 01, 10, 11, 000, 001, 010, 011, 100, 101, \ldots .$$

Note that natural numbers can be represented as binary words, e.g., by means of the binary representation: $\text{bin}(n)$ denotes the binary representation of n. Now decision problems for numbers can be expressed as sets of words over $\{0, 1\}$, e.g.

$$\text{SQUARE} = \{\text{bin}(n) \mid n \geq 0 \text{ is a square}\}$$
$$= \{0, 1, 100, 1001, 10000, 11001, 100100, 110001, 1000000, \ldots\}$$

codes the problem "Given n, decide whether n is a square of some number", while

PRIMES = {bin(n) | $n \geq 2$ is a prime number}

= {10, 11, 101, 111, 1011, 1101, 10001, 10011, 10111, 11101, ...}

codes the primality problem. Every subset of Σ^* is called a **language**. Thus, SQUARE and PRIMES are languages.

In computability and complexity theory, algorithms for inputs that are words over a finite alphabet are traditionally formalized as programs for a particular machine model, the Turing machine. Readers who are interested in the formal details of measuring the time complexity of algorithms in terms of this model are referred to standard texts on computational complexity theory such as [23]. Basically, the model charges one step for performing one operation involving a fixed number of letters, or digits.

We say that a language $L \subseteq \Sigma^*$ is in class P if there is a Turing machine (program) M and a polynomial p such that on input $x \in \Sigma^*$ consisting of m letters the machine M makes no more than $p(m)$ steps and arrives at the answer 1 if $x \in L$ and at the answer 0 if $x \notin L$.

For our purposes it is sufficient to note the following: if we have an algorithm $\mathcal{A}$ that operates on numbers (like Algorithms 1.1.1 and 1.2.1) so that the total number of operations that are performed on input n is bounded by $c(\log n)^k$ for constants c and k and so that the intermediate results never become larger than n^k, then the language

{bin(n) | $\mathcal{A}$ on input n outputs 0}

is in class P.

Thus, to establish that PRIMES is in P it is sufficient to find an algorithm $\mathcal{A}$ for the primality problem that operates on (not too large) numbers with a polynomial operation bound. The question of whether such an algorithm might exist had been open ever since the terminology for asking the question was developed in the 1960s.

1.4 Randomized and Superpolynomial Time Algorithms for the Primality Problem

In a certain sense, the search for polynomial time algorithms for the primality problem was already successful in the 1970s, when two very efficient methods for testing large numbers for primality were proposed, one by Solovay and Strassen [39], and one by Rabin [35], based on previous work by Miller [29]. These algorithm have the common feature that they employ a random experiment (just like Algorithm 1.2.1); so they fall into the category of *randomized algorithms*. For both these algorithms the following holds.

- If the input is a prime number, the output is 0.
- If the input is composite, the output is 0 or 1, and the probability that the outcome is 0 is bounded by $\frac{1}{2}$.

On input n, both algorithms use at most $c \cdot \log n$ arithmetic operations on numbers not larger than n^2, for some constant c; i.e., they are about as fast as Algorithm 1.2.1. If the output is 1, the input number n is definitely composite; we say that the calculation *proves* that n is composite, and yields a certificate for that fact. If the result is 0, we do not really know whether the input number is prime or not. Certainly, an error bound of up to $\frac{1}{2}$ is not satisfying. However, by repeating the algorithm up to ℓ times, hence spending $\ell \cdot c \cdot \log n$ arithmetic operations on input n, the error bound can be reduced to $2^{-\ell}$, for arbitrary ℓ. And if we choose to carry out $\ell = d \log n$ repetitions, the algorithms will still have a polynomial operation bound, but the error bound drops to n^{-d}, extremely small for n with a hundred or more decimal digits.

These randomized algorithms, along with others with a similar behavior (e.g., the Lucas Test and the Frobenius Test, described in [16, Sect. 3.5]), are sufficient for solving the primality problem for quite large inputs for all practical purposes, and algorithms of this type are heavily used in practice. For practical purposes, there is no reason to worry about the risk of giving output 0 on a composite input n, as long as the error bound is adjusted so that the probability for this to happen is smaller than 1 in 10^{20}, say, and it is guaranteed that the algorithm exhibits the behavior as if truly random coin tosses were available. Such a small error probability is negligible in relation to other (hardware or software) error risks that are inevitable with real computer systems. The Miller-Rabin Test and the Solovay-Strassen Test are explored in detail later in this text (Chaps. 5 and 6).

Still, from a theoretical point of view, the question remained whether there was an absolutely error-free algorithm for solving the primality problem with a small time bound. Here one may consider

(a) algorithms without randomization (called **deterministic** algorithms to emphasize the contrast), and
(b) randomized algorithms with *expected* polynomial running time which never give erroneous outputs.

As for (a), the (up to the year 2002) fastest known deterministic algorithm for the primality problem was proposed in 1983 by Adleman, Pomerance, and Rumeley [2]. It has a time bound of $f_7^c(\log n)$, where $f_7^c(x) = x^{c \ln \ln x}$ for some constant $c > 0$, which makes it slightly superpolynomial. Practical implementations have turned out to be successful for numbers with many hundreds of decimal digits [12].

As for (b), in 1987 Adleman and Huang [1] proposed a randomized algorithm that has a (high-degree) polynomial time bound and yields primality certificates, in the following sense: On input n, the algorithm outputs 0 or 1. If the output is 1, the input n is guaranteed to be prime, and the calculation carried out by the algorithm constitutes a proof of this fact. If the input n is a prime number, then the probability that the wrong answer 0 is given is at most $\frac{1}{n}$. Algorithms with this kind of behavior are called ***primality proving*** algorithms.

The algorithm of Adleman and Huang ($\mathcal{A}_{AH}$) may be combined with, for example, the Solovay-Strassen Test ($\mathcal{A}_{SS}$) to obtain an error-free randomized algorithm for the primality problem with expected polynomial time bound, as follows: Given an input n, run *both* algorithms on n. If one of them gives a definite answer ($\mathcal{A}_{AH}$ declares that n is a prime number or $\mathcal{A}_{SS}$ declares that n is composite), we are done. Otherwise, keep repeating the procedure until an answer is obtained. The expected number of repetitions is smaller than 2 no matter whether n is prime or composite. The combined algorithm gives the correct answer with probability 1, and the ***expected*** time bound is polynomial in $\log n$.

There are further algorithms that provide proofs for the primality of an input number n, many of them quite successful in practice. For much more information on primality testing and primality proving algorithms see [16]. (A complete list of the known algorithms as of 2004 may be found in the overview paper [11].)

1.5 The New Algorithm

Such was the state of affairs when in August 2002 M. Agrawal, N. Kayal, and N. Saxena published their paper "PRIMES is in P". In this paper, Agrawal, Kayal, and Saxena described a deterministic algorithm for the primality problem, and a polynomial bound of $c \cdot (\log n)^{12} \cdot (\log \log n)^d$ was proved for the number of bit operations, for constants c and d.

In the time analysis of the algorithm, a deep result of Fouvry [19] from analytical number theory was used, published in 1985. This result concerns the density of primes of a special kind among the natural numbers. Unfortunately, the proof of Fouvry's theorem is accessible only to readers with a quite strong background in number theory. In discussions following the publication of the new algorithm, some improvements were suggested. One of these improvements (by H.W. Lenstra [10, 27]) leads to a slightly modified algorithm with a new time analysis, which avoids the use of Fouvry's theorem altogether, and makes it possible to carry out the time analysis and correctness proof solely by basic methods from number theory and algebra. The new analysis even yields an improved bound of $c \cdot (\log n)^{10.5} \cdot (\log \log n)^d$ on the number of bit operations. Employing Fouvry's result one obtains the even smaller bound $c \cdot (\log n)^{7.5} \cdot (\log \log n)^d$.

Experiments and number-theoretical conjectures make it seem likely that the exponent in the complexity bound can be chosen even smaller, about 6 instead of 7.5. The reader may consult Table 1.1 to get an idea for numbers of which order of magnitude the algorithm is guaranteed to terminate in reasonable time. Currently, improvements of the new algorithm are being investigated, and these may at some time make it competitive with the primality proving algorithms currently in use. (See [11].)

Citing the title of a review of the result [13], with the improved and simplified time analysis the algorithm by Agrawal, Kayal, and Saxena appears even more a "Breakthrough for Everyman": a result that can be explained to interested high-school students, with a correctness proof and time analysis that can be understood by everyone with a basic mathematical training as acquired in the first year of studying mathematics or computer science. It is the purpose of this text to describe this amazing and impressive result in a self-contained manner, along with two randomized algorithms (Solovay-Strassen and Miller-Rabin) to represent practically important primality tests.

The book covers just enough material from basic number theory and elementary algebra to carry through the analysis of these algorithms, and so frees the reader from collecting methods and facts from different sources.

1.6 Finding Primes and Factoring Integers

In cryptographic applications, e.g., in the RSA cryptosystem [36], we need to be able to solve the **prime generation** problem, i.e., produce multidigit randomly chosen prime numbers. Given a primality testing algorithm $\mathcal{A}$ with one-sided error, like the Miller-Rabin Test (Chap. 5), one may generate a random prime in $[10^s, 10^{s+1} - 1]$ as follows: Choose an odd number a from this interval at random; run $\mathcal{A}$ on a. If the outcome indicates that a is prime, output a, otherwise start anew with a new random number a.

For this algorithm to succeed we need to have some information about the density of prime numbers in $[10^s, 10^{s+1}-1]$. It is a consequence of Chebychev's Theorem 3.6.3 below that the fraction of prime numbers in $[10^s, 10^{s+1} - 1]$ exceeds c/s, for some constant $c > 0$. This implies that the number of trials needed until the randomly chosen number a is indeed a prime number is no larger than s/c. The expected computation cost for obtaining an output is then no larger than s/c times the cost of running algorithm $\mathcal{A}$. If the probability that algorithm $\mathcal{A}$ declares a composite number a prime is no larger than $2^{-\ell}$, then the probability that the output is composite is no larger than $2^{-\ell} \cdot s/c$, which can be made as small as desired by choosing ℓ large enough. We see that the complexity of generating primes is tightly coupled with the complexity of primality testing. In practice, thus, the advent of the primality test of Agrawal, Kayal, and Saxena has not changed much with respect to the problem of generating primes, since it is much slower than the randomized algorithms and the error probability can be made so small that it is irrelevant from the point of view of the applications.

On the other hand, for the security of many cryptographic systems it is important that the **factoring problem**

Given a composite number n, find a proper factor of n

is *not* easily solvable for n sufficiently large. An introduction into the subject of factoring is given in, for example, [41]; an in-depth treatment may be found in [16]. As an example, we mention one algorithm from the family of the fastest known factorization algorithms, the "number field sieve", which has a superpolynomial running time bound of $c \cdot e^{d \cdot (\ln n)^{1/3} (\ln \ln n)^{2/3}}$, for a constant d a little smaller than 1.95 and some $c > 0$. Using algorithms like this, one has been able to factor single numbers of more than 200 decimal digits.

It should be noted that with respect to factoring (and to the security of cryptosystems that are based on the supposed difficulty of factoring) no change is to be expected as a consequence of the new primality test. This algorithm shares with all other fast primality tests the property that if it declares an input number n composite, in most cases it does so on the basis of indirect evidence, having detected a property in n prime numbers cannot have. Such a property usually does not help in finding a proper factor of n.

1.7 How to Read This Book

Of course, the book may be read from cover to cover. In this way, the reader is lead on a guided tour through the basics of algorithms for numbers, of number theory, and of algebra (including all the proofs), as far as they are needed for the analysis of the three primality tests treated here.

Chapter 2 should be checked for algorithmic notation and basic algorithms for numbers. Readers with some background in basic number theory and/or algebra may want to read Sects. 3.1 through 3.5 and Sects. 4.1 through 4.3 only cursorily to make sure they are familiar with the (standard) topics treated there. Section 3.6 on the density bounds for prime numbers and Sect. 4.4 on the fact that in finite fields the multiplicative group is cyclic are a little more special and provide essential building blocks of the analysis of the new primality test by Agrawal, Kayal, and Saxena.

Chapters 5 and 6 treat the Miller-Rabin Test and the Solovay-Strassen Test in a self-contained manner; a proof of the quadratic reciprocity law, which is used for the time analysis of the latter algorithm, is provided in Appendix A.3. These two chapters may be skipped by readers interested exclusively in the deterministic primality test.

Chapter 7 treats polynomials, in particular polynomials over finite fields and the technique of constructing finite fields by quotienting modulo an irreducible polynomial. Some special properties of the polynomial $X^r - 1$ are developed there. All results compiled in this section are essential for the analysis of the deterministic primality test, which is given in Chap. 8.

Readers are invited to send information about mistakes, other suggestions for improvements, or comments directly to the author's email address:

`martin.dietzfelbinger@tu-ilmenau.de`

A list of corrections will be held on the webpage
http://eiche.theoinf.tu-ilmenau.de/kt/pbook

2. Algorithms for Numbers and Their Complexity

The notion of an algorithm is basic in computer science. Usually, one says that an algorithm is a finite piece of text that describes in an unambiguous way which elementary computational steps are to be performed on any given input, and in which way the result should be read off after the computation has ended. In the theory of algorithms and in computational complexity theory, one traditionally formalizes the notion of an algorithm as a program for a particular theoretical machine model, the Turing machine. In our context, where we deal with numbers rather than with strings, this is not appropriate, hence we use a different notation for algorithms, described in Sect. 2.1. As a technical prerequisite for discussing complexity issues, we introduce O-notation in Sect. 2.2. In Sect. 2.3 the complexity of some elementary operations on numbers is discussed.

2.1 Notation for Algorithms on Numbers

We describe algorithms in an informal framework ("pseudocode"), resembling imperative programs for simplified computers with one CPU and a main memory. For readers without programming experience we briefly describe the main features of the notation.

We use only two elementary data structures:

- A *variable* may contain an integer. (Variables are denoted by typewriter type names like a, b, k, l, and so on.)
- An *array* corresponds to a sequence of variables, indexed by a segment $\{1, \ldots, k\}$ of the natural numbers. (Arrays are denoted by typewriter letters, together with their index range in square brackets; an array element is given by the name with the index in brackets. Thus, a[1..100] denotes an array with 100 components; a[37] is the 37th component of this array.) Each array component may contain an integer.

If not obvious from the context, we list the variables and arrays used in an algorithm at the beginning. In many algorithms numbers are used that do not change during the execution; such numbers, so-called *constants*, are denoted by the usual mathematical notation.

M. Dietzfelbinger: Primality Testing in Polynomial Time, LNCS 3000, pp. 13-21, 2004.
© Springer-Verlag Berlin Heidelberg 2004

There are two basic ways in which constants, variables, and array components can be used:

- *Usage*: write the name of the variable to extract and use its content by itself or in an expression. Similarly, constants are used by their name.
- *Assigning a new value*: if v is some integral value, e.g., obtained by evaluating some expression, then $x \leftarrow v$ is an instruction that causes this value to be put into x.

By combining variables and constants from $\mathbb{Z}$ with operations like addition and multiplication, parenthesized if necessary, we form expressions, which are meant to cause the corresponding computation to be carried out. For example, the instruction

$$x[i] \leftarrow (a + b) \text{ div } c$$

causes the current contents of a and b to be added and the result to be divided by the constant c. The resulting number is stored in component $x[i]$, where i is the number currently stored in variable i.

By comparing the values of numerical expressions with a comparison operator from $\{\leq, \geq, <, >, =, \neq\}$, we obtain *boolean values* from $\{true, false\}$, which may further be combined using the usual boolean operators in $\{\wedge, \vee, \neg\}$, to yield *boolean expressions*.

Further elementary instructions are the **return** statement that immediately finishes the execution of the algorithm, and the **break** statement that causes a loop to be finished.

Next, we describe ways in which elementary instructions (assignments, **return**, **break**) may be combined to form more complicated program segments or *statements*. Formally, this is done using an inductive definition. Elementary instructions are statements. If $stm_1, \ldots, stm_r$, $r \geq 1$, are statements, then the sequence $\{stm_1; \cdots stm_r; \}$ is also a statement; the semantics is that the statements $stm_1, \ldots, stm_r$ are to be carried out one after the other. In our notation for algorithms, we use an indentation scheme to avoid curly braces: a consecutive sequence of statements that are indented to the same depth is to be thought of as being enclosed in curly braces.

Further, we use **if-then** statements and **if-then-else** statements with the obvious semantics. For *bool_expr* an expression that evaluates to a boolean value and *stm* a statement, the statement

$$\textbf{if } bool\_expr \textbf{ then } stm$$

is executed as follows: first the boolean expression is evaluated to some value in $\{true, false\}$; if and only if the result is *true*, the (simple or composite) statement *stm* is carried out. Similarly, a statement

$$\textbf{if } bool\_expr \textbf{ then } stm_1 \textbf{ else } stm_2$$

is executed as follows: if the value of the boolean expression is *true*, then stm_1 is carried out, otherwise stm_2 is carried out.

In order to be able to write repetitive instructions in a concise way, we use *for loops*, *while loops*, and *repeat loops*. A **for** statement

$$\textbf{for } \texttt{i} \textbf{ from } expr_1 \textbf{ to } expr_2 \textbf{ do } stm$$

has the following semantics. The expressions $expr_1$ and $expr_2$ are evaluated, with integral results n_1 and n_2. Then the "loop body" *ins* is executed $\max\{0, n_2 - n_1 + 1\}$ times, once with i containing n_1, once with i containing $n_1 + 1$, ..., once with i containing n_2. Finally i is assigned the value $n_2 + 1$ (or n_1, if $n_1 > n_2$). It is understood that the loop body does not contain an assignment for i. The loop body may contain the special instruction **break**, which, when executed, immediately terminates the execution of the loop, without changing the contents of i. Again, we use indentation to indicate how far the body of the loop extends. If instead of the keyword **to** we use **downto**, then the content of i is decreased in each execution of the loop body.

The number of repetitions of a **while** loop is not calculated beforehand. Such a statement, written

$$\textbf{while } bool\_expr \textbf{ do } stm$$

has the following semantics. The boolean expression $bool\_expr$ is evaluated. If the outcome is *true*, the body *stm* is carried out once, and we start again carrying out the whole **while** statement. Otherwise the execution of the statement is finished.

A **repeat** statement is similar. It has the syntax

$$\textbf{repeat } stm \textbf{ until } bool\_expr$$

and is executed as follows. The statement *stm* is carried out once. Then the boolean expression $bool\_expr$ is evaluated. If the result is *true*, the execution of the statement is finished. Otherwise we start again carrying out the whole **repeat** statement. Just as with a **for** loop, execution of a **while** loop or a **repeat** loop may also be finished by executing a **break** statement.

A special elementary instruction is the operation **random**. The statement **random**(r), where $r \geq 2$ is a number, returns as value a randomly chosen element of $\{0, \ldots, r - 1\}$. If this instruction is used in an algorithm, the result becomes a random quantity and the running time might become a random variable. If the **random** instruction is used in an algorithm, we call it a *"randomized algorithm"*.

2.2 *O*-notation

In algorithm analysis, it is convenient to have a notation for running times which is not clogged up by too much details, because often it is not possible and often it would even be undesirable to have an exact formula for the number of operations. We want to be able to express that the running time grows at most "at the rate of $\log n$" with the input n, or that the number of bit operations needed for a certain task is roughly proportional to $(\log n)^3$. For this, "O-notation" (read "big-Oh-notation") is commonly used.

We give the definitions and basic rules for manipulating bounds given in O-notation. The proofs of the formulas and claims are simple exercises in calculus. For more details on O-notation the reader may consult any text on the analysis of algorithms (e.g., [15]).

Definition 2.2.1. *For a function* $f\colon \mathbb{N} \to \mathbb{R}^+$ *we let*

$$O(f) = \{g \mid g\colon \mathbb{N} \to \mathbb{R}^+, \exists C > 0 \exists n_0 \forall n \geq n_0 : g(n) \leq C \cdot f(n)\},$$
$$\Omega(f) = \{g \mid g\colon \mathbb{N} \to \mathbb{R}^+, \exists c > 0 \exists n_0 \forall n \geq n_0 : g(n) \geq c \cdot f(n)\},$$
$$\Theta(f) = O(f) \cap \Omega(f).$$

Alternatively, we may describe $O(f)$ [or $\Omega(f)$ or $\Theta(f)$, resp.] as the set of functions g with $\limsup_{n\to\infty} \frac{g(n)}{f(n)} < \infty$ [or $\liminf_{n\to\infty} \frac{g(n)}{f(n)} > 0$, or both, resp.].

Example 2.2.2. Let $f_i(n) = (\log n)^i$, for $i = 1, 2, \ldots$.

(a) If g is a function with $g(n) \leq 50(\log n)^2 - 100(\log\log n)^2$ for all $n \geq 100$, then $g \in O(f_2)$, and $g \in O(f_i)$ for all $i > 2$.
(b) If g is a function with $g(n) \geq (\log n)^3 \cdot (\log\log n)^2$ for all $n \geq 50$, then $g \in \Omega(f_3)$ and $g \in \Omega(f_2)$.
(c) If g is a function with $\frac{1}{10}\log n \leq g(n) \leq 15\log n + 5(\log\log n)^2$ for all $n \geq 50$, then $g \in \Theta(f_1)$.

Thus, O-notation helps us in classifying functions g by simple representative growth functions f even if the exact values of the functions g are not known. Note that since we demand a certain behavior only for $n \geq n_0$, it does not matter if the functions that we consider are not defined or not specified for some initial values $n < n_0$.

Usually, one writes:

$$g(n) = O(f(n)) \quad \text{if } g \in O(f),\,^1$$
$$g(n) = \Omega(f(n)) \quad \text{if } g \in \Omega(f), \text{ and}$$
$$g(n) = \Theta(f(n)) \quad \text{if } g \in \Theta(f)\,.$$

Sometimes, we use $O(f(n))$ also as an abbreviation for an arbitrary function in $O(f)$; e.g., we might say that algorithm $\mathcal{A}$ has a running time of $O(f(n))$ if the running time $t_{\mathcal{A}}(n)$ on input n satisfies $t_{\mathcal{A}}(n) = O(f(n))$. Extending this, we write $O(g(n)) = O(f(n))$ if *every* function in $O(g)$ is also in $O(f)$.

---

[1] Read: "$g(n)$ is big-Oh of $f(n)$".

The reader should be aware that the equality sign is not used in a proper way here; in particular, the relation $O(g(n)) = O(f(n))$ is not symmetric.

In a slight extension of this convention, we write $g(n) = O(1)$ if there is a constant C such that $g(n) \leq C$ for all n, and $g(n) = \Omega(1)$ if $g(n) \geq c$ for all n, for some $c > 0$.

For example, if the running time $t_{\mathcal{A}}(n)$ of some algorithm $\mathcal{A}$ on input n is bounded above by $50(\log n)^2 - 10(\log \log n)^2$, we write $t_{\mathcal{A}}(n) = O((\log n)^2)$. Similarly, if $c \leq t_{\mathcal{B}}(n)/(\log n)^3 < C$ for all $n \geq n_0$, we write $t_{\mathcal{B}}(n) = \Theta((\log n)^3)$.

There are some simple rules for combining estimates in O-notation, which we will use without further comment:

Lemma 2.2.3. (a) *if $g_1(n) = O(f_1(n))$ and $g_2(n) = O(f_2(n))$, then*
$g_1(n) + g_2(n) = O(\max\{f_1(n), f_2(n)\})$;
(b) *if $g_1(n) = O(f_1(n))$ and $g_2(n) = O(f_2(n))$, then*
$g_1(n) \cdot g_2(n) = O(f_1(n) \cdot f_2(n))$. □

For example, if $g_1(n) = O((\log n)^2)$ and $g_2(n) = O((\log n)(\log \log n))$, then $g_1(n) + g_2(n) = O((\log n)^2)$ and $g_1(n) \cdot g_2(n) = O((\log n)^3(\log \log n))$.

We extend the O-notation to functions of two variables and write

$$g(n, m) = O(f(n, m))$$

if there is a constant $C > 0$ and there are n_0 and m_0 such that for all $n \geq n_0$ and $m \geq m_0$ we have $g(n, m) \leq C \cdot f(n, m)$. Of course, this may be generalized to more than two variables. Using this notation, we note another rule that is essential in analyzing the running time of loops. It allows us to use summation of $O(\ldots)$-terms.

Lemma 2.2.4. *If $g(n, i) = O(f(n, i))$, and $G(n, m) = \sum_{1 \leq i \leq m} g(n, i)$ and $F(n, m) = \sum_{1 \leq i \leq m} f(n, i)$, then $G(n, m) = O(F(n, m))$.* □

A particular class of time bounds are the polynomials (in $\log n$): Note that if $g(n) \leq a_d(\log n)^d + \cdots + a_1 \log n + a_0$ for arbitrary integers $a_d, \ldots, a_1, a_0$, then $g(n) = O((\log n)^d)$.

We note that if $\lim_{n \to \infty} \frac{f_1(n)}{f_2(n)} = 0$ then $O(f_1) \subsetneq O(f_2)$. For example, $O(\log n) \subsetneq O((\log n)(\log \log n))$. The following sequence of functions characterize more and more extensive O-classes:

$$\log n, (\log n)(\log \log n), (\log n)^{3/2}, (\log n)^2, (\log n)^2(\log \log n)^3, (\log n)^{5/2},$$
$$(\log n)^3, (\log n)^4, \ldots, (\log n)^{\log \log \log n} = 2^{(\log \log n)(\log \log \log n)},$$
$$2^{(\log \log n)^2}, 2^{\sqrt{\log n}}, 2^{(\log n)/2} = \sqrt{n}, n^{3/4}, n/\log n, n, n \log \log n, n \log n.$$

In connection with the bit complexity of operations on numbers, like primality tests, one often is satisfied with a coarser classification of complexity bounds that ignores logarithmic factors.

Definition 2.2.5. *For a function* $f\colon \mathbb{N} \to \mathbb{R}^+$ *with* $\lim_{n\to\infty} f(n) = \infty$ *we let*

$$O^\sim(f) = \{g \mid g\colon \mathbb{N} \to \mathbb{R}^+, \exists C > 0 \exists n_0 \exists k \forall n \geq n_0 : g(n) \leq C \cdot f(n) \log(f(n))^k\}.$$

Again, we write $g(n) = O^\sim(f(n))$ if $g \in O^\sim(f)$. A typical example is the bit complexity $t(n) = O((\log n)(\log \log n)(\log \log \log n))$ of the Schönhage-Strassen method for multiplying integers n and $m \leq n$ [41]. This complexity is classified as $t(n) = O^\sim(\log n)$. Likewise, when discussing a time bound $(\log n)^{7.5}(\log \log n)^4 (\log \log \log n)$ one prefers to concentrate on the most significant factor and write $O^\sim((\log n)^{7.5})$ instead. The rules from Lemma 2.2.3 also apply to the "$O^\sim$-notation".

2.3 Complexity of Basic Operations on Numbers

In this section, we review the complexities of some elementary operations with integers.

For a natural number $n \geq 1$, let $\|n\| = \lceil \log_2(n+1) \rceil$ be the number of bits in the binary representation of n; let $\|0\| = 1$. We recall some simple bounds on the number of *bit operations* it takes to perform arithmetic operations on natural numbers given in binary representation. The formulas would be the same if decimal notation were used, and multiplication and addition or subtraction of numbers in $\{0, 1, \ldots, 9\}$ were taken as basic operations. In all cases, the number of operations is polynomial in the bit length of the input numbers.

Fact 2.3.1. *Let* n, m *be natural numbers.*

(a) *Adding or subtracting* n *and* m *takes* $O(\|n\| + \|m\|) = O(\log n + \log m)$ *bit operations.*

(b) *Multiplying* m *and* n *takes* $O(\|n\| \cdot \|m\|) = O(\log(n) \cdot \log(m))$ *bit operations.*

(c) *Computing the quotient* n div m *and the remainder* $n \bmod m$ *takes* $O((\|n\| - \|m\| + 1) \cdot \|m\|)$ *bit operations.*[2]

Proof. In all cases, the methods taught in school, adapted to binary notation, yield the claimed bounds.   □

Addition and subtraction have cost linear in the binary length of the input numbers, which is optimal. The simple methods for multiplication and division have cost no more than quadratic in the input length. For our main theme of polynomial time algorithms this is good enough. However, we note that much faster methods are known.

---

[2] The operations div and mod are defined formally in Definition 3.1.9.

Fact 2.3.2. *Assume n and m are natural numbers of at most k bits each.*

(a) *([38]) We may multiply m and n with $O(k(\log k)(\log\log k)) = O^\sim(k)$ bit operations.*

(b) *(For example, see [41].) We may compute n div m and n mod m using $O(k(\log k)(\log\log k)) = O^\sim(k)$ bit operations.* $\square$

An interesting and important example for an elementary operation on numbers is modular exponentiation. Assume we wish to calculate the number $2^{10987654321} \bmod 101$. Clearly, the naive way — carrying out 10987654320 multiplications and divisions modulo 101 — is inefficient if feasible at all. Indeed, if we calculated $a^n \bmod m$ by doing $n-1$ multiplications and divisions by m, the number of steps needed would grow exponentially in the bit length of the input, which is $\|a\| + \|n\| + \|m\|$.

There is a simple, but very effective trick to speed up this operation, called "repeated squaring". The basic idea is that we may calculate the powers $s_i = a^{2^i} \bmod m$, $i \geq 0$, by the recursive formula

$$s_0 = a \bmod n; \quad s_i = s_{i-1}^2 \bmod m, \text{ for } i \geq 1.$$

Thus, k multiplications and divisions by m are sufficient to calculate $a^{2^i} \bmod m$, for $0 \leq i \leq k$. Further, if $b_k b_{k-1} \cdots b_1 b_0$ is the binary representation of n, i.e., $n = \sum_{0 \leq i \leq k, b_i = 1} 2^i$, then

$$a^n \bmod m = \left(\prod_{\substack{0 \leq i \leq k \\ b_i = 1}} s_i \right) \bmod m.$$

This means that at most k further multiplications and divisions are sufficient to calculate $a^n \bmod m$. Let us see how this idea works by carrying out the calculation for $2^{4321} \bmod 101$ (Table 2.1). We precalculate that the binary representation of 4321 is $b_k \cdots b_0 = 1000011100001$. By c_i we denote the partial product $\prod_{0 \leq j \leq i, b_j = 1} a^{2^j} \bmod m$. It only remains to efficiently obtain the bits of the binary representation of n. Here the trivial observation helps that the last bit b_0 is 0 or 1 according as n is even or odd, and that in the case $n \geq 2$ the preceding bits $b_k \cdots b_1$ are just the binary representation of $\lfloor n/2 \rfloor = a$ div 2. Thus, a simple iterative procedure yields these bits in the order $b_0, b_1, \ldots, b_k$ needed by the method sketched above. If we interleave both calculations, we arrive at the following procedure.

Algorithm 2.3.3 (Fast Modular Exponentiation)

INPUT: Integers a, n, and $m \geq 1$.
METHOD:
```
0      u, s, c: integer;
1      u ← n;
```

```
2       s ← a mod m;
3       c ← 1;
4       while u ≥ 1 repeat
5           if u is odd then c ← (c · s) mod m;
6           s ← s · s mod m;
7           u ← u div 2;
8       return c;
```

i	s_i	b_i	c_i
0	$2^1 = 2$	1	2
1	$2^2 = 4$	0	2
2	$4^2 = 16$	0	2
3	$16^2 \bmod 101 = 54$	0	2
4	$54^2 \bmod 101 = 88$	0	2
5	$88^2 \bmod 101 = 68$	1	$2 \cdot 68 \bmod 101 = 35$
6	$68^2 \bmod 101 = 79$	1	$35 \cdot 79 \bmod 101 = 38$
7	$79^2 \bmod 101 = 80$	1	$38 \cdot 80 \bmod 101 = 10$
8	$80^2 \bmod 101 = 37$	0	10
9	$37^2 \bmod 101 = 56$	0	10
10	$56^2 \bmod 101 = 5$	0	10
11	$5^2 \bmod 101 = 25$	0	10
12	$25^2 \bmod 101 = 19$	1	$10 \cdot 19 \bmod 101 = 89$

Table 2.1. Exponentiation by repeated squaring

In line 2 s is initialized with $a \bmod m = a^{2^0} \bmod m$, and u with n. In each iteration of the loop, u is halved in line 7; thus when line 5 is executed for the $(i+1)$st time, u contains the number with binary representation $b_k b_{k-1} \ldots b_i$. Further, in line 6 of each iteration, the contents of s are squared (modulo m); thus when line 5 is executed for the $(i+1)$st time, s contains $s_i = a^{2^i} \bmod m$. The last bit b_i of u is used in line 5 to decide whether p should be multiplied by s_i or not. This entails, by induction, that after carrying out line 5 the ith time p contains $\prod_{0 \le j \le i} a^{2^j} \bmod m$. The loop stops when u has become 0, which obviously happens after $k+1$ iterations have been carried out. At this point p contains $\prod_{0 \le j \le k} a^{2^j} \bmod m$, the desired result.

Lemma 2.3.4. *Calculating $a^n \bmod m$ takes $O(\log n)$ multiplications and divisions of numbers from $\{0, \ldots, m^2 - 1\}$, and $O((\log n)(\log m)^2)$ (naive) resp. $O^\sim((\log n)(\log m))$ (advanced methods) bit operations.* □

It is convenient to here provide an algorithm for testing natural numbers for the property of being a perfect power. We say that $n \ge 1$ is a *perfect power* if $n = a^b$ for some $a, b \ge 2$. Obviously, only exponents b with $b \le \log n$

can satisfy this. The idea is to carry out the following calculation for *each*
such b: We may check for any given $a < n$ whether $a^b < n$ or $a^b = n$ or
$a^b > n$. (Using the Fast Exponentiation Algorithm 2.3.3 without modulus,
but cutting off as soon as an intermediate result larger than n appears will
keep the numbers to be handled smaller than n^2.) This makes it possible to
conduct a *binary search* in $\{1, \ldots, n\}$ for a number a that satisfies $a^b = n$.

In detail, the algorithm looks as follows:

Algorithm 2.3.5 (Perfect Power Test)

INPUT: Integer $n \geq 2$.
METHOD:

```
0      a, b, c, m: integer;
1      b ← 2;
2      while 2^b ≤ n repeat
3          a ← 1; c ← n;
4          while c − a ≥ 2 repeat
5              m ← (a + c) div 2;
6              p ← min{m^b, n + 1};    (∗ fast exponentiation, truncated ∗)
7              if p = n then return ("perfect power", m, b);
8              if p < n then a ← m else c ← m ;
9          b ← b + 1;
10     return "no perfect power";
```

In lines 3–8 a binary search is carried out, maintaining the invariant that
for the contents a, c of a, c we always have $a^b < n < c^b$. In a round, the
median $m = (a + c)$ div 2 is calculated and (using the fast exponentiation
algorithm) the power m^b is calculated; however, as soon as numbers larger
than n appear in the calculation, we break off and report the answer $n+1$. In
this way, numbers larger than n never appear as factors in a multiplication.
If and when $m = (a + c)$ div 2 satisfies $m^b = n$, the algorithm stops and
reports success (line 7). Otherwise either a or c is updated to the new value
m, so that the invariant is maintained. In each round through the loop 3–8
the number of elements in the interval $[a+1, c-1]$ halves; thus, the loop runs
for at most $\log n$ times before $c - b$ becomes 0 or 1. The outer loop (lines 2–9)
checks all possible exponents, of which there are $\lfloor \log n \rfloor - 1$ many. Summing
up, we obtain:

Lemma 2.3.6. *Testing whether n is a perfect power is not more expen-*
sive than $O((\log n)^2 \log \log n)$ multiplications of numbers from $\{1, \ldots, n\}$.
This can be achieved with $O((\log n)^4 \log \log n)$ bit operations (naive) or even
$O^{\sim}((\log n)^3)$ bit operations (fast multiplication).   □

We remark that using a different approach ("Newton iteration") algorithms
for perfect power testing may be devised that need only $O^{\sim}((\log n)^2)$ bit
operations. (See [41, Sect. 9.6].)

3. Fundamentals from Number Theory

In this chapter, we study those notions from number theory that are essential for divisibility problems and for the primality problem. It is important to understand the notion of greatest common divisors and the Euclidean Algorithm, which calculates greatest common divisors, and its variants. The Euclidean Algorithm is the epitome of efficiency among the number-theoretical algorithms. Further, we introduce modular arithmetic, which is basic for all primality tests considered here. The Chinese Remainder Theorem is a basic tool in the analysis of the randomized primality tests. Some important properties of prime numbers are studied, in particular the unique factorization theorem. Finally, both as a general background and as a basis for the analysis of the deterministic primality test, some theorems on the density of prime numbers in the natural numbers are proved.

3.1 Divisibility and Greatest Common Divisor

The actual object of our studies in this book are the **natural numbers** $0, 1, 2, 3, \ldots$ and the **prime numbers** $2, 3, 5, 7, 11, \ldots$. Still, it is necessary to start with considering the set

$$\mathbb{Z} = \{0, 1, -1, 2, -2, 3, -3, \ldots\}$$

of **integers** together with the standard operations of addition, subtraction, and multiplication. This structure is algebraically much more convenient than the natural numbers.

Definition 3.1.1. *An integer n **divides** an integer m, in symbols $n \mid m$, if $nx = m$ for some integer x. If this is the case we also say that n is a **divisor** of m or that m is a **multiple** of n.*

Example 3.1.2. Every integer is a multiple of 1 and of -1; the multiples of 3 are $0, 3, -3, 6, -6, 9, -9, \ldots$. The number 0 does not divide anything excepting 0 itself. On the other hand, 0 is a multiple of *every* integer.

We list some elementary properties of the divisibility relation, which will be used without further comment later.

M. Dietzfelbinger: Primality Testing in Polynomial Time, LNCS 3000, pp. 23-53, 2004.
© Springer-Verlag Berlin Heidelberg 2004

Proposition 3.1.3. (a) *If $n \mid m$ and $n \mid k$, then $n \mid (mu + kv)$ for arbitrary integers u and v.*
(b) *If $n, m > 0$ and $n \mid m$, then $n \leq m$.*
(c) *If $n \mid m$ and $m \mid k$, then $n \mid k$.*
(d) *If $n \mid m$, then $(-n) \mid m$ and $n \mid (-m)$.*

Proof. (a) If $m = nx$ and $k = ny$, then $mu + kv = n(xu + yv)$.
(b) Assume $m = nx$. Then $x > 0$, which implies $x \geq 1$, since x is an integer. Thus, $m - n = nx - n = n \cdot (x - 1) \geq 0$.
(c) If $m = nx$ and $k = my$, then $k = n(xy)$.
(d) If $m = nx$, then $m = (-n)(-x)$ and $-m = n(-x)$.   □
    Because of (d) in the preceding proposition, it is enough to study nonnegative divisors.

Definition 3.1.4. *For an integer n, let $D(n)$ denote the set of nonnegative divisors of n.*

Example 3.1.5. (a) $D(0)$ comprises all nonnegative integers.
(b) $D(60) = D(-60) = \{1, 2, 3, 4, 5, 6, 10, 12, 15, 20, 30, 60\}$.

For $n \neq 0$, $D(n)$ contains only positive numbers, with 1 among them. By Proposition 3.1.3 we have that $D(n) = D(-n)$, that $D(n)$ is partially ordered by the relation $\mid$, and that it is downward closed under this relation.

Definition 3.1.6. (a) *If n and m are integers, not both 0, then the largest integer that divides n and divides m is called the **greatest common divisor of n and** m and denoted by $\gcd(n, m)$.*
    (Note that the set $D(n) \cap D(m)$ of common divisors of n and m contains 1, hence is nonempty, and that it is finite, hence has a maximum.)
    It is convenient to define (somewhat arbitrarily) $\gcd(0, 0) = 0$.
(b) *If $\gcd(n, m) = 1$, then we say that n and m are **relatively prime**.*

For example, we have $\gcd(4, 15) = 1$, $\gcd(4, 6) = 2$, $\gcd(4, 0) = \gcd(4, 4) = 4$. The motivation for the term *relatively prime* will become clearer below when we treat prime numbers. We note some simple rules for manipulating $\gcd(\cdot, \cdot)$ values.

Proposition 3.1.7. (a) $\gcd(n, m) = \gcd(-n, m) = \gcd(n, -m)$
    $= \gcd(-n, -m)$, *for all n, m. In particular, $\gcd(n, m) = \gcd(|n|, |m|)$.*
(b) $\gcd(n, n) = \gcd(n, 0) = \gcd(0, n) = |n|$ *for all integers n.*
(c) $\gcd(n, m) = \gcd(m, n)$, *for all integers n, m.*
(d) $\gcd(n, m) = \gcd(n + mx, m)$, *for all integers n, m, x.*

Proof. (a) holds because $D(n) = D(-n)$ for all integers n.
(b) and (c) are immediate from the definitions.
(d) If $m = 0$, the claim is trivially true. Thus, assume $m \neq 0$, hence $D(m)$ is a finite set. Obviously then it is sufficient to note that $D(n) \cap D(m) = D(n + mx) \cap D(m)$. This is proved as follows:

"⊆": If $n = du$ and $m = dv$, then $n + mx = d(u + vx)$, hence $n + mx$ is a multiple of d as well.

"⊇": If $n + mx = du$ and $m = dv$ then $n = d(u - vx)$, hence n is a multiple of d as well.   □

Another important rule for $\gcd(\cdot, \cdot)$ calculations will be given in Corollary 3.1.12.

Proposition 3.1.8 (Integer Division with Remainder).   *If n is an integer and d is a positive integer (the **divisor** or **modulus**), then there are unique integers q (the **quotient**) and r (the **remainder**) such that*

$$n = dq + r \quad and \quad 0 \le r < d.$$

There are quite efficient algorithms for finding such numbers q and r from n and d, discussed in Sect. 2.3. Here, we prove existence and uniqueness (indicating a very inefficient algorithm for the task of finding q and r).

Proof. Existence: Let q be the maximal integer such that $qd \le a$. (Since $(-|n|)d \le n < (|n| + 1)d$, this q is well defined and can in principle be found by searching through a finite set of integers.) The choice of q implies that $qd \le n < (q+1)d$. We define $r = n - qd$, and conclude $0 \le r < d$, as desired. *Uniqueness*: Assume $n = qd + r = q'd + r'$, for $0 \le r, r' < d$. By symmetry, we may assume that $r' - r \ge 0$; obviously, then, $0 \le r' - r < d$. Hence $0 = (q' - q)d + (r' - r)$, which means that $r' - r$ is a multiple of d. By Proposition 3.1.3(b), the only multiple of d in $\{0, 1, \ldots, d-1\}$ is 0, hence we must have $r = r'$. This in turn implies $qd = q'd$, from which we get $q = q'$, since d is nonzero.   □

The operation of division with remainder is so important that we introduce special notation for it.

Definition 3.1.9. *For an integer n and a positive integer d we let*

$$n \bmod d = r, \quad and$$
$$n \operatorname{div} d = q,$$

where r and q are the uniquely determined numbers that satisfy $n = qd + r$, $0 \le r < d$, as in Proposition 3.1.8.

Note that $n \operatorname{div} d = \lfloor n/d \rfloor$, which denotes the largest integer not exceeding n/d. (See Appendix A.2.)

Examples:

$$
\begin{aligned}
16 \operatorname{div} 4 &= \quad 4, & 16 \bmod 4 &= 0, \\
9 \operatorname{div} 4 &= \quad 2, & 9 \bmod 4 &= 1, \\
3 \operatorname{div} 4 &= \quad 0, & 3 \bmod 4 &= 3, \\
0 \operatorname{div} 4 &= \quad 0, & 0 \bmod 4 &= 0, \\
-2 \operatorname{div} 4 &= -1, & -2 \bmod 4 &= 2, \\
-9 \operatorname{div} 4 &= -3, & -9 \bmod 4 &= 3.
\end{aligned}
$$

The following property of division with remainder is essential for the Euclidean Algorithm, an efficient method for calculating the greatest common divisor of two numbers (see Sect. 3.2).

Proposition 3.1.10. *If $m \geq 1$, then $\gcd(n, m) = \gcd(n \bmod m, m)$ for all integers n.*

Proof. Since $n \bmod m = n - qm$ for $q = n$ div m, this is a special case of Proposition 3.1.7(d). □

The greatest common divisor of n and m has an important property, which is basic for many arguments and constructions in this book: it is an integral linear combination of n and m.

Proposition 3.1.11. *For arbitrary integers n and m there are integers x and y such that*

$$\gcd(n, m) = nx + my.$$

Proof. If $n = 0$ or $m = 0$, either $(x, y) = (1, 1)$ or $(x, y) = (-1, -1)$ will be suitable. So assume both n and m are nonzero. Let $I = \{nu + mv \mid u, v \in \mathbb{Z}\}$. Then I has positive elements, e.g., $n^2 + m^2$ is a positive element of I. Choose x and y so that $d = nx + my > 0$ is the *smallest* positive element in I. We claim that $d = \gcd(n, m)$.

For this we must show that
(i) $d \in D(n) \cap D(m)$, and that
(ii) all positive elements of $D(n) \cap D(m)$ divide d.
Assertion (ii) is simple: use that if k divides n and m, then k divides nx and my, hence also the sum $d = nx + my$. To prove (i), we show that d divides n; the proof for m is the same. By Proposition 3.1.8, we may write $n = dq + r$ for some integer q and some r with $0 \leq r < d$. Thus,

$$r = n - dq = n - (nx + my)q = n(1 - xq) + m(-yq),$$

which shows that $r \in I$. Now d was chosen to be the smallest positive element of I, and $r < d$, so it must be the case that $r = 0$, which implies that $n = dq$. Thus d is a divisor of n, as desired. □

Corollary 3.1.12. *For all integers n and m, and $k > 0$ we have*

$$\gcd(kn, km) = k \cdot \gcd(n, m).$$

Proof. If $n = m = 0$, there is nothing to show. Thus, assume that $d = \gcd(n, m) > 0$, and write $d = nx + my$ for integers x and y. Then $kd = (kn)x + (km)y$, hence every common divisor of kn and km divides kd. On the other hand, it is clear that kd divides both kn and km. Thus, $kd = \gcd(kn, km)$. □

The following most important special case of Proposition 3.1.11 will be used without further comment later.

Proposition 3.1.13. *For integers n and m the following are equivalent:*

(i) *n and m are relatively prime, and*

(ii) *there are integers x and y so that $1 = nx + my$.*

Proof. (i) $\Rightarrow$ (ii) is just Proposition 3.1.11 for $1 = \gcd(n, m)$. For the direction (ii) $\Rightarrow$ (i), note that if $1 = nx + my$, then n and m cannot be both 0, and every common divisor of n and m also divides 1, hence $\gcd(n, m) = 1$.
□

For example, for $n = 20$ and $m = 33$ we have $1 = 33 \cdot (-3) + 20 \cdot 5 = 33 \cdot 17 + 20 \cdot (-28)$. More generally, if $nx + my = 1$, then clearly $n(x + um) + m(y - un) = 1$ for arbitrary $u \in \mathbb{Z}$.

We note two consequences of Proposition 3.1.13.

Corollary 3.1.14. *For all integers n, m, and k we have: If n and k are relatively prime, then $\gcd(n, mk) = \gcd(n, m)$.*

Proof. Since n and k are relatively prime, we can write $1 = nx + ky$ for suitable integers x, y. This implies $m = n(mx) + (mk)y$, from which it is immediate that every common divisor of n and mk also divides m. Thus, $D(n) \cap D(m) = D(n) \cap D(mk)$, which implies the claim.
□

Proposition 3.1.15. *If n and m are relatively prime integers, and n and m both divide k, then nm divides k.*

Proof. Assume $k = ns$ and $k = mt$, for integers s, t. By Proposition 3.1.13 we may write $1 = nx + my$ for integers x and y. Then

$$k = k \cdot nx + k \cdot my = mt \cdot nx + ns \cdot my = (tx + sy) \cdot nm,$$

which proves the claim.
□

Continuing the example just mentioned, let us take the number 7920, which equals $20 \cdot 396$ and $33 \cdot 240$. Using the argument from the previous proof, we see that $7920 = (396 \cdot (-3) + 240 \cdot 5) \cdot (20 \cdot 33) = 12 \cdot (20 \cdot 33)$.

3.2 The Euclidean Algorithm

The Euclidean Algorithm is a cornerstone in the area of number-theoretic algorithms. It provides an extremely efficient method for calculating the greatest common divisor of two natural numbers. An extended version even calculates a representation of the greatest common divisor of n and m as a linear combination of n and m (see Proposition 3.1.11). The algorithm is based on the repeated application of the rule noted as Proposition 3.1.10. We start with the classical Euclidean Algorithm, formulated in the simplest way. (There are other formulations, notably ones that use recursion.)

Algorithm 3.2.1 (Euclidean Algorithm)

INPUT: Two integers n, m.

METHOD:

```
0       a, b: integer;
1       if |n| ≥ |m|
2           then a ← |n|; b ← |m|;
3           else b ← |m|; a ← |n|;
4       while b > 0 repeat
5           (a, b) ← (b, a mod b);
6       return a;
```

In lines 1–3 the absolute values of the input numbers n and m are placed into the variables a and b in such a way that b is not larger than a. Lines 4 and 5 form a loop. In each iteration of the loop the remainder $a \bmod b$ is computed and placed into b (as the divisor in the next iteration); simultaneously, the old value of b is put into a. In this way, the algorithm generates a sequence

$$(a_0, b_0), (a_1, b_1), \ldots, (a_t, b_t),$$

where $\{a_0, b_0\} = \{|n|, |m|\}$ and $a_i = b_{i-1}$, $b_i = a_{i-1} \bmod b_{i-1}$, for $1 \leq i \leq t$, and $b_t = 0$.

For an example, consider Table 3.1, in which is listed the sequence of numbers that the algorithm generates on input $n = 10534$, $m = 12742$. The numbers a_i and b_i are the contents of variables a and b after the loop has been executed i times. The value returned is 46. We will see in a moment that this

i	a_i	b_i
0	12742	10534
1	10534	2208
2	2208	1702
3	1702	506
4	506	184
5	184	138
6	138	46
7	46	0

Table 3.1. The Euclidean Algorithm on $n = 10534$ and $m = 12742$

means that the greatest common divisor of 10536 and 12742 is 46. Indeed, it is not hard to see that the Euclidean Algorithm always returns the value $\gcd(n, m)$, for integers n and m. To see this, we prove a "loop invariant".

Lemma 3.2.2. (a) *For the pair (a_i, b_i) stored in a and b after the loop in lines 4 and 5 has been carried out for the ith time we have*

$$\gcd(a_i, b_i) = \gcd(n, m). \tag{3.2.1}$$

(b) *On input n, m, Algorithm 3.2.1 returns the greatest common divisor of m and n. If both m and n are 0, the value returned is also 0.*

Proof. (a) We use induction on i. By Proposition 3.1.7(a) and (c), $\gcd(a_0, b_0)$ $= \gcd(n, m)$, since $\{a_0, b_0\} = \{|n|, |m|\}$. Further, for $i \geq 1$,

$$\gcd(a_i, b_i) = \gcd(b_{i-1}, a_{i-1} \bmod b_{i-1}) = \gcd(a_{i-1}, b_{i-1}) = \gcd(n, m),$$

by the instruction in line 5, Proposition 3.1.10, and the induction hypothesis. (b) It is obvious that in case $n = m = 0$ the loop is never performed and the value returned is 0. Thus assume that n and m are not both 0. To see that the loop terminates, it is sufficient to observe that $b_0, b_1, b_2, \ldots$ is a strictly decreasing sequence of integers, hence there must be some t with $b_t = 0$. That means that after finitely many executions of the loop variable b will get the value 0, and the loop terminates. For the content a_t of a at this point, which is the returned value, we have $a_t = \gcd(a_t, 0) = \gcd(n, m)$, by part (a) and Proposition 3.1.7(b).   □

Next, we analyze the complexity of the Euclidean Algorithm. It will turn out that it has a running time linear in the number of bits of the input numbers (in terms of arithmetic operations) and quadratic cost (in terms of bit operations). In fact, the cost is not more than that of multiplying m and n (in binary) by the naive method. This means that on a computer the Euclidean Algorithm can be carried out very quickly for numbers that have hundreds of bits, and in reasonable time even if they have thousands of bits.

Lemma 3.2.3. *Assume Algorithm 3.2.1 is run on input n, m. Then we have:*

(a) *The loop in lines 4 and 5 is carried out at most $2\min\{\|n\|, \|m\|\} = O(\min\{\log(n), \log(m)\})$ times.*
(b) *The number of bit operations made is $O(\log(n)\log(m))$.*

Proof. (a) We have already seen in the previous proof that the numbers $b_0, b_1, \ldots$ form a strictly decreasing sequence. A closer look reveals that the decrease is quite fast. Consider three subsequent values b_i, b_{i+1}, b_{i+2}.

Case 1: $b_{i+1} > \frac{1}{2}b_i = \frac{1}{2}a_{i+1}$. Then $b_{i+2} = a_{i+1} - b_{i+1} < \frac{1}{2}b_i$.
Case 2: $b_{i+1} \leq \frac{1}{2}b_i = \frac{1}{2}a_{i+1}$. Then $b_{i+2} = a_{i+1} \bmod b_{i+1} < b_{i+1} \leq \frac{1}{2}b_i$.

This means that in two rounds the bit length of the content of variable b is reduced by 1 (unless it has reached the value 0 anyway). Thus, after at most $2\min\{\|n\|, \|m\|\}$ executions of the loop b contains 0, and the loop stops.

(b) Even if we use the naive method for dividing a_i by b_i (in binary notation), $O((\|a_i\| - \|b_i\| + 1) \cdot \|b_i\|)$ bit operations are sufficient for this operation, see Fact 2.3.1(c). Note that $b_i = a_{i+1}$, for $0 \leq i < t$, and hence

$$(\|a_i\| - \|b_i\| + 1)\|b_i\| = \|a_i\| \cdot \|b_i\| - \|a_{i+1}\| \cdot \|b_i\| + \|b_i\| \leq (\|a_i\| - \|a_{i+1}\|)\|b_0\| + \|b_i\|.$$

Thus, the total number of bit operations needed in lines 4 and 5 can be bounded by

$$\sum_{0 \le i < t} O((\|a_i\| - \|b_i\| + 1)\|b_i\|)$$

$$= O\Big(\sum_{0 \le i < t} ((\|a_i\| - \|a_{i+1}\|)\|b_0\| + \|b_i\|) \Big)$$

$$= O(\|a_0\| \cdot \|b_0\| + t\|b_0\|) = O(\|n\| \cdot \|m\|).$$

The comparison in line 1 of the algorithm takes $O(\|n\| + \|m\|)$ bit operations. Thus the overall cost is $O(\|n\| \cdot \|m\|) = O(\log(n) \log(m))$, as claimed. $\square$

We now turn to an extended version of the Euclidean Algorithm. We have noted in Proposition 3.1.11 that the greatest common divisor d of n and m can be written as

$$d = nx + my,$$

for certain integers x and y. In our example from Table 3.1, we can write

$$46 = 12742 \cdot (-62) + 10534 \cdot 75 = 12742 \cdot 167 + 10534 \cdot (-202),$$

as is easily checked using a pocket calculator. Actually, there are infinitely many such pairs x, y, since if $d = nx + my$, then obviously we also have $d = n(x + k(m/d)) + m(y - k(n/d))$, for arbitrary $k \in \mathbb{Z}$. But how to find the first such pair? Slightly extending the Euclidean Algorithm helps.

Algorithm 3.2.4 (Extended Euclidean Algorithm)

INPUT: Two integers n and m.
METHOD:
```
0      a, b, xa, ya, xb, yb: integer;
1      if |n| ≥ |m|
2          then a ← |n|; b ← |m|;
3                xa ← sign(n); ya ← 0; xb ← 0; yb ← sign(m);
4          else a ← |m|; b ← |n|;
5                xa ← 0; ya ← sign(m); xb ← sign(n); yb ← 0;
6      while b > 0 repeat
7          q ← a div b;
8          (a, b) ← (b, a − q · b);
9          (xa, ya, xb, yb) ← (xb, yb, xa − q · xb, ya − q · yb);
10     return (a, xa, ya);
```

In the algorithm we use the signum function, defined by

$$\mathrm{sign}(n) = \begin{cases} 1 & , \text{ if } n > 0, \\ 0 & , \text{ if } n = 0, \\ -1 & , \text{ if } n < 0, \end{cases} \qquad (3.2.2)$$

with the basic property that

$$n = \mathrm{sign}(n) \cdot |n| \quad \text{for all } n. \qquad (3.2.3)$$

We note that with respect to the variables a and b nothing has changed in comparison to the original Euclidean Algorithm, since $a - q \cdot b$ is the same as a mod b. But an additional quadruple of numbers is carried along in variables xa, ya, xb, yb. These variables are initialized in lines 3 and 5 and change in parallel with a and b in the body of the loop. Finally, in line 10 the contents of xa and ya are returned along with $\gcd(n, m)$. We want to see that these two numbers are the coefficients we are looking for.

Let $x_{a,i}$, $y_{a,i}$, $x_{b,i}$, $y_{b,i}$ denote the contents of the variables xa, ya, xb, yb after the loop in lines 6–9 has been carried out i times. Table 3.2 gives the numbers obtained in the course of the computation if Algorithm 3.2.4 is applied to the same numbers as in the example from Table 3.1. For completeness, also the quotients $q_i = a_{i-1}$ div b_{i-1}, $i = 1, \ldots, 7$, are listed.

i	a_i	b_i	$x_{a,i}$	$y_{a,i}$	$x_{b,i}$	$y_{b,i}$	q_i
0	12742	10534	0	1	1	0	–
1	10534	2208	1	0	−1	1	1
2	2208	1702	−1	1	5	−4	4
3	1702	506	5	−4	−6	5	1
4	506	184	−6	5	23	−19	3
5	184	138	23	−19	−52	43	2
6	138	46	−52	43	75	−62	1
7	46	0	75	−62	−277	229	3

Table 3.2. The Extended Euclidean Algorithm on $n = 10534$ and $m = 12742$

The output is the triple $(46, 75, -62)$. We have already noted that 75 and -62 are coefficients that may be used for writing $d = \gcd(n, m)$ as a linear combination of n and m. The following lemma states that the result of the Extended Euclidean Algorithm always has this property.

Lemma 3.2.5. *If on input n, m the extended Euclidean Algorithm outputs (d, x, y), then $d = \gcd(n, m) = nx + my$.*

Proof. We have seen before that when the algorithm terminates after t executions of the loop, variable a contains $d = \gcd(n, m)$. For the other part of the claim, we prove by induction on i that for all $i \leq t$ we have

$$a_i = nx_{a,i} + my_{a,i} \text{ and } b_i = nx_{b,i} + my_{b,i}. \tag{3.2.4}$$

For $i = 0$, (3.2.4) holds by (3.2.3) and the way the variables are initialized in lines 1–5. For the induction step, assume (3.2.4) holds for $i - 1$. Then we have, by the way the variables are updated in the ith iteration of the loop:

$$a_i = b_{i-1} = nx_{b,i-1} + my_{b,i-1} = nx_{a,i} + my_{a,i},$$

and

$$b_i = a_{i-1} - q_i b_{i-1} = nx_{a,i-1} + my_{a,i-1} - q_i(nx_{b,i-1} + my_{b,i-1}) = nx_{b,i} + my_{b,i}.$$

In particular, the coefficients $x_{a,t}$ and $y_{a,t}$ stored in xa and ya after the last iteration of the loop satisfy $\gcd(n, m) = a_t = nx_{a,t} + my_{a,t}$, as claimed.   □

Concerning the running time of the Extended Euclidean Algorithm, we note that the analysis in Lemma 3.2.3(a) carries over, so on input n, m $O(\min\{\log(n), \log(m)\})$ arithmetic operations are carried out. As for the cost in terms of bit operations, we note without proof that the number of bit operations is bounded by $O(\log(n)\log(m))$ just as in the case of the simple Euclidean Algorithm.

3.3 Modular Arithmetic

We now turn to a different view on remainders: ***modular arithmetic***. Let $m \geq 2$ be given (the "***modulus***"). We want to say that looking at an integer a we are not really interested in a but only in the remainder $a \bmod m$. Thus, all numbers that leave the same remainder when divided by m are considered "similar". We define a binary relation on $\mathbb{Z}$.

Definition 3.3.1. *Let $m \geq 2$ be given. For arbitrary integers a and b we say that a is **congruent to** b **modulo** m and write*

$$a \equiv b \pmod{m}$$

if $a \bmod m = b \bmod m$.

The definition immediately implies the following properties of the binary relation "congruence modulo m".

Lemma 3.3.2. *Congruence modulo m is an **equivalence relation**, i.e., we have*
Reflexivity: $a \equiv a \pmod{m}$,
Symmetry: $a \equiv b \pmod{m}$ *implies* $b \equiv a \pmod{m}$, *and*
Transitivity: $a \equiv b \pmod{m}$ *and* $b \equiv c \pmod{m}$ *implies* $a \equiv c \pmod{m}$.   □

Further, it is almost immediate from the definitions of $a \bmod m$ and of $\equiv$ that

$$a \equiv b \pmod{m} \quad \text{if and only if} \quad m \text{ divides } b - a. \tag{3.3.5}$$

(Write $a = mq + r$ and $b = mq' + r'$ with $0 \leq r, r' < m$. Then $b - a = m(q' - q) + (r' - r)$ is divisible by m if and only if $r' - r$ is divisible by m. But $|r' - r| < m$, so the latter is equivalent to $r = r'$.) Property (3.3.5) is often used as the definition of the congruence relation. It is important to note that this relation is compatible with the arithmetic operations on $\mathbb{Z}$ (which fact is expressed by using the word "congruence relation").

Lemma 3.3.3. *If $a \equiv a'$ (mod m) and $b \equiv b'$ (mod m), then $a + b \equiv a' + b'$ (mod m) and $a \cdot b \equiv a' \cdot b'$ (mod m), for all $a, a', b, b' \in \mathbb{Z}$. Consequently, if $a \equiv a'$ and $n \geq 0$ is arbitrary, then $a^n \equiv (a')^n$ (mod m).*

Proof. As an example, we consider the multiplicative rule. Write $a' = a + qm$ and $b' = b + rm$. Then $a' \cdot b' = a \cdot b + (qb + ar + qrm)m$, which implies that $a \cdot b \equiv a' \cdot b'$ (mod m). □

In many cases, this lemma makes calculating a remainder $f(a_1, \ldots, a_r)$ mod m easier, for $f(x_1, \ldots, x_r)$ an arbitrary arithmetic expression. We will use it without further comment by freely substituting equivalent terms in calculations. To demonstrate the power of these rules, consider the task of calculating the remainder $(751^{100} - 22^{59})$ mod 4. Using Lemma 3.3.3, we see (since $751 \equiv 3$, $3^2 \equiv 1$, $22 \equiv 2$, and $2^2 \equiv 0$, all modulo 4):

$$751^{100} - 22^{59} \equiv 3^{100} - 2 \cdot (2^2)^{29} \equiv (3^2)^{50} - 2 \cdot 0 \equiv 1^{50} \equiv 1 \quad (\text{mod } 4),$$

and hence $(751^{100} - 22^{59})$ mod $4 = 1$.

Like all equivalence relations, congruence modulo m splits its ground set $\mathbb{Z}$ into equivalence classes (or ***congruence classes***). There is exactly one equivalence class for each remainder $r \in \{0, 1, \ldots, m-1\}$, since a is congruent to $a \bmod m \in \{0, 1, \ldots, m-1\}$ and distinct $r, r' \in \{0, 1, \ldots, m-1\}$ cannot be congruent. For $m = 4$ these equivalence classes are:

$$\{a \in \mathbb{Z} \mid a \bmod 4 = 0\} = \{ \ldots, -12, -8, -4, 0, 4, \;\; 8, 12, \ldots \},$$
$$\{a \in \mathbb{Z} \mid a \bmod 4 = 1\} = \{ \ldots, -11, -7, -3, 1, 5, \;\; 9, 13, \ldots \},$$
$$\{a \in \mathbb{Z} \mid a \bmod 4 = 2\} = \{ \ldots, -10, -6, -2, 2, 6, 10, 14, \ldots \},$$
$$\{a \in \mathbb{Z} \mid a \bmod 4 = 3\} = \{ \ldots, \;\; -9, -5, -1, 3, 7, 11, 15, \ldots \}.$$

We introduce an arithmetic structure on these classes. For convenience, we use the standard representatives from $\{0, 1, \ldots, m-1\}$ as names for the classes, and calculate with these representatives.

Definition 3.3.4. *For $m \geq 2$ let $\mathbb{Z}_m$ be the set $\{0, 1, \ldots, m-1\}$. On this set the following two operations $+_m$ (**addition modulo** m) and $\cdot_m$ (**multiplication modulo** m) are defined:*

$$a +_m b = (a + b) \bmod m \quad and \quad a \cdot_m b = (a \cdot b) \bmod m.$$

(The subscript m at the operation symbols is omitted if no confusion arises.)

The operations $+_m$ and $\cdot_m$ obey the standard arithmetic laws known from the integers: associativity, commutativity, distributivity. Moreover, both operations have neutral elements, and $+_m$ has inverses.

Lemma 3.3.5. (a) $a +_m b = b +_m a$ *and* $a \cdot_m b = b \cdot_m a$, *for* $a, b \in \mathbb{Z}_m$.
(b) $(a +_m b) +_m c = a +_m (b +_m c)$ *and* $(a \cdot_m b) \cdot_m c = a \cdot_m (b \cdot_m c)$, *for* $a, b, c \in \mathbb{Z}_m$.

(c) $(a +_m b) \cdot_m c = a \cdot_m c +_m b \cdot_m c$, for $a, b, c \in \mathbb{Z}_m$.
(d) $a +_m 0 = 0 +_m a = a$ and $a \cdot_m 1 = 1 \cdot_m a = a$, for $a \in \mathbb{Z}_m$.
(e) $a +_m (m - a) = (m - a) +_m a = 0$, for $a \in \mathbb{Z}_m$.

Proof. The proofs of these rules all follow the same pattern, namely one shows that the expressions involved are congruent modulo m, and then concludes that their remainders are equal. For example, the distributivity law (c) is proved as follows: Using Lemma 3.3.3 and the fact that always $(a \bmod m) \equiv a$ $(\bmod m)$, we get

$$(a +_m b) \cdot c = ((a + b) \bmod m) \cdot c \equiv (a + b)c \quad (\bmod m)$$

and

$$a \cdot_m c + b \cdot_m c = (ac \bmod m) + (bc \bmod m) \equiv ac + bc \equiv (a + b)c \quad (\bmod m).$$

By transitivity, $(a +_m b) \cdot c \equiv a \cdot_m c + b \cdot_m c \pmod m$, hence $(a +_m b) \cdot_m c = ((a +_m b) \cdot c) \bmod m = (a \cdot_m c + b \cdot_m c) \bmod m = a \cdot_m c +_m b \cdot_m c$. □
    Since $0 \cdot_m a = 0$ for all a, the element 0 is uninteresting when looking at multiplication modulo m. Very often, the set $\mathbb{Z}_m - \{0\}$ is not a "nice" structure with respect to multiplication modulo m, since it is not closed under this operation. For example, $12 \cdot_{18} 9 = 108 \bmod 18 = 0$. More generally, if $d = \gcd(a, m) > 1$, then $b = \frac{m}{\gcd(a,m)} < m$ and $a \cdot_m b = \frac{a}{\gcd(a,m)} \cdot m \bmod m = 0$. But note the following cancellation rule.

Proposition 3.3.6 (Cancellation Rule).

(a) *If* $m \mid ab$ *and* $\gcd(m, a) = 1$, *then* m *divides* b.
(b) *If* $ab \equiv ac \pmod m$ *and* $\gcd(m, a) = 1$, *then* $b \equiv c \pmod m$.

Proof. (a) We write $1 = mx + ay$. Then $b = m(bx) + (ab)y$. Since m divides ab by assumption, a must divide b.
(b) Assume that $ab \equiv ac \pmod m$, i.e., $a(b - c) \equiv 0 \pmod m$. This means that m divides $a(b - c)$. By (a), we conclude that m divides $b - c$, that means $b \equiv c \pmod m$. □
    Elements a with $\gcd(a, m) = 1$ play a special role in $\mathbb{Z}_m$, because on this set the operation $\cdot_m$ behaves nicely.

Definition 3.3.7. *For* $m \geq 1$ *let*

$$\mathbb{Z}_m^* = \{a \mid 1 \leq a < m, \ \gcd(a, m) = 1\} \quad and \quad \varphi(m) = |\mathbb{Z}_m^*|.$$

The function φ is called ***Euler's φ-function*** or ***Euler's totient function***.

Proposition 3.3.8. (a) $1 \in \mathbb{Z}_m^*$.
(b) *If* $a, b \in \mathbb{Z}_m^*$, *then* $a \cdot_m b \in \mathbb{Z}_m^*$.
(c) $a \in \mathbb{Z}_m^*$ *if and only if there is some* $b \in \mathbb{Z}_m$ *with* $a \cdot_m b = 1$.

Proof. (a) is trivial. — For (b) and (c) we make heavy use of Proposition 3.1.13:

(b) Assume $\gcd(a, m) = \gcd(b, m) = 1$. We write

$$1 = ax + my \quad \text{and} \quad 1 = bu + mv,$$

for integers x, y, u, v. Then

$$(ab) \cdot (xu) = (ax)(bu) = (1 - my) \cdot (1 - mv) = 1 - m(y + v - myv),$$

hence $\gcd(ab, m) = 1$. By Proposition 3.1.10, this implies $\gcd(ab \bmod m, m) = 1$.

(c) "$\Rightarrow$": Assume $\gcd(a, m) = 1$. Then there are integers x and y with

$$1 = ax + my.$$

This implies $ax - 1 = -my \equiv 0 \pmod{m}$. Let $b = x \bmod m$. Then $ab \equiv ax \equiv 1 \pmod{m}$, which implies that $a \cdot_m b = 1$.

"$\Leftarrow$": Assume $ab \bmod m = 1$. This means that $ab - 1 = mx$ for some x, which implies $\gcd(a, m) = 1$. $\qquad\Box$

Note that by the formula in the proof of (c) "$\Rightarrow$" it is implied that for $a \in \mathbb{Z}_m^*$ some b with $a \cdot_m b = 1$ can be found efficiently by applying the Extended Euclidean Algorithm 3.2.4 to a and m.

Example 3.3.9. (a) In $\mathbb{Z}_7^* = \{1, 2, 3, 4, 5, 6\}$ we have $3 \cdot_7 3 = 2$ and $3 \cdot_7 5 = 1$. By inspection, $\varphi(7) = 6$.

(b) In $\mathbb{Z}_{15}^* = \{1, 2, 4, 7, 8, 11, 13, 14\}$ the products $4 \cdot_{15} 1, 4 \cdot_{15} 2, 4 \cdot_{15} 4, \ldots, 4 \cdot_{15} 14$ of 4 with elements of $\mathbb{Z}_{15}^*$ are $4, 8, 1, 13, 2, 14, 7, 11$. By inspection, $\varphi(15) = 8$.

(c) In $\mathbb{Z}_{27}^* = \{1, 2, 4, 5, 7, 8, 10, 11, 13, 14, 16, 17, 19, 20, 22, 23, 25, 26\}$ we have $8 \cdot_{27} 17 = 1$. By inspection, $\varphi(27) = 18$.

In Sect. 3.5 we will see how to calculate $\varphi(n)$ for numbers n whose prime decomposition is known.

3.4 The Chinese Remainder Theorem

We start with an example. Consider $24 = 3 \cdot 8$, a product of two numbers that are relatively prime. We set up a table of the remainders $a \bmod 3$ and $a \bmod 8$, for $0 \le a < 24$. (Note that if $a \in \mathbb{Z}$ is arbitrary, then $a \bmod 3 = (a \bmod 24) \bmod 3$, so the remainders modulo 3 (and 8) of other numbers are obtained by cyclically extending the table.)

If in Table 3.3 we consider the entries in rows 2 and 3 and rows 5 and 6 as 24 pairs in $\mathbb{Z}_3 \times \mathbb{Z}_8$, we observe that these are all different, hence cover all 24 possibilities in $\{0, 1, 2\} \times \{0, 1, \ldots, 7\}$. Thus the mapping $a \mapsto (a \bmod 3, a \bmod 8)$ is a bijection between $\mathbb{Z}_{24}$ and $\mathbb{Z}_3 \times \mathbb{Z}_8$. But more is true: arithmetic operations carried out with elements of $\{0, 1, \ldots, 23\}$ are

a	0	1	2	3	4	5	6	7	8	9	10	11
$a \bmod 3$	0	1	2	0	1	2	0	1	2	0	1	2
$a \bmod 8$	0	1	2	3	4	5	6	7	0	1	2	3

a	12	13	14	15	16	17	18	19	20	21	22	23
$a \bmod 3$	0	1	2	0	1	2	0	1	2	0	1	2
$a \bmod 8$	4	5	6	7	0	1	2	3	4	5	6	7

Table 3.3. Remainders of $0, 1, \ldots, 23$ modulo 3 and modulo 8

mirrored in the remainders modulo 3 and 8. For example, addition of the pairs $(2, 7)$ and $(2, 1)$ yields $(1, 0)$, which corresponds to adding 23 and 17 to obtain $40 \equiv 16 \pmod{24}$. Similarly,

$$(2^5 \bmod 3, 3^5 \bmod 8) = (2, 3),$$

which corresponds to the observation that $11^5 \bmod 24 = 5$.

The Chinese Remainder Theorem says in essence that such a structural connection between the remainders modulo n and pairs of remainders modulo n_1, n_2 will hold whenever $n = n_1 n_2$ for n_1, n_2 relatively prime.

Theorem 3.4.1. *Let $n = n_1 n_2$ for n_1, n_2 relatively prime. Then the mapping*

$$\Phi \colon \mathbb{Z}_n \to \mathbb{Z}_{n_1} \times \mathbb{Z}_{n_2}, \; a \mapsto (a \bmod n_1, a \bmod n_2)$$

is a bijection. Moreover, if $\Phi(a) = (a_1, a_2)$ and $\Phi(b) = (b_1, b_2)$, then

(a) $\Phi(a +_n b) = (a_1 +_{n_1} b_1, a_2 +_{n_2} b_2)$;
(b) $\Phi(a \cdot_n b) = (a_1 \cdot_{n_1} b_1, a_2 \cdot_{n_2} b_2)$;
(c) $\Phi(a^m \bmod n) = ((a_1)^m \bmod n_1, (a_2)^m \bmod n_2)$, *for $m \geq 0$.*

Proof. We show that Φ is one-to-one. (Since $|\mathbb{Z}_n| = n = n_1 n_2 = |\mathbb{Z}_{n_1} \times \mathbb{Z}_{n_2}|$, this implies that Φ is a bijection.) Thus, assume $0 \leq a \leq b < n$ and $\Phi(a) = \Phi(b)$, i.e.,

$$a \bmod n_1 = b \bmod n_1 \quad \text{and} \quad a \bmod n_2 = b \bmod n_2.$$

We rewrite this as

$$(b - a) \bmod n_1 = 0 \quad \text{and} \quad (b - a) \bmod n_2 = 0,$$

i.e., $b - a$ is divisible by n_1 and by n_2. Since n_1 and n_2 are relatively prime, we conclude by Proposition 3.1.15 that n divides $b - a$. Since $0 \leq b - a < n$, this implies that $a = b$. Altogether this means that Φ is one-to-one, as claimed.

As for the rules (a)–(c), we only look at part (b) and show that

$$\Phi(a \cdot_n b) = (a_1 \cdot_{n_1} b_1, a_2 \cdot_{n_2} b_2). \tag{3.4.6}$$

(Parts (a) and (c) are proved similarly.) By the assumptions, we have

$$a \equiv a_1 \bmod n_1 \quad \text{and} \quad b \equiv b_1 \bmod n_1.$$

By Lemma 3.3.3, this implies $a \cdot b \equiv a_1 \cdot b_1 \pmod{n_1}$. On the other hand, since n_1 divides n, we have $a \cdot b \equiv ((a \cdot b) \bmod n) = a \cdot_n b \pmod{n_1}$. By transitivity, we get $(a \cdot_n b) \bmod n_1 = a_1 \cdot_{n_1} b_1$, which is one half of (3.4.6). The other half, concerning n_2, a_2, b_2, is proved in the same way.   □

Theorem 3.4.1 may be interpreted as to say that calculating modulo n is equivalent to calculating "componentwise" modulo n_1 and n_2. Further, we will often use it in the following form. Prescribing the remainders modulo n_1 and n_2 determines uniquely a number in $\{0, \ldots, n-1\}$.

Corollary 3.4.2. *If $n = n_1 n_2$ for n_1, n_2 relatively prime, then for arbitrary integers x_1 and x_2 there is exactly one $a \in \mathbb{Z}_n$ with*

$$a \equiv x_1 \pmod{n_1} \quad \text{and} \quad a \equiv x_2 \pmod{n_2}. \tag{3.4.7}$$

Proof. Define $a_1 = x_1 \bmod n_1$ and $a_2 = x_2 \bmod n_2$. By Theorem 3.4.1 there is exactly one $a \in \mathbb{Z}_n$ such that

$$a \equiv a_1 \pmod{n_1} \quad \text{and} \quad a \equiv x_2 \pmod{n_2}. \tag{3.4.8}$$

This a is as required. Uniqueness follows from the fact that all solutions a to (3.4.7) must satisfy (3.4.8), and that this is unique by Theorem 3.4.1.   □

The Chinese Remainder Theorem can be generalized to an arbitrarily large finite number of factors.

Theorem 3.4.3. *Let $n = n_1 \cdots n_r$ for $n_1, \ldots, n_r$ relatively prime. Then the mapping*

$$\Phi \colon \mathbb{Z}_n \to \mathbb{Z}_{n_1} \times \cdots \times \mathbb{Z}_{n_r}, \quad a \mapsto (a \bmod n_1, \ldots, a \bmod n_r)$$

is a bijection, with isomorphism properties (a)–(c) analogous to those formulated in Theorem 3.4.1.

Proof. We just indicate the proof of the claim that Φ is a bijection. Since

$$n = |\mathbb{Z}_n| = n = n_1 \cdots n_r = |\mathbb{Z}_{n_1} \times \cdots \times \mathbb{Z}_{n_r}|,$$

again it is sufficient to show that Φ is one-to-one. Assume $0 \le a \le b < n$ and

$$(a \bmod n_1, \ldots, a \bmod n_r) = (b \bmod n_1, \ldots, b \bmod n_r).$$

Then $b - a$ is divisible by $n_1, \ldots, n_r$, and since these numbers are relatively prime, $b - a$ is also divisible by n. Since $0 \le b - a < n$, we conclude $a = b$.

The isomorphism properties are easily checked, just as in the case with two factors.   □

It is an interesting and important consequence of the Chinese Remainder Theorem that Φ also provides a bijection between $\mathbb{Z}_n^*$ and $\mathbb{Z}_{n_1}^* \times \mathbb{Z}_{n_2}^*$. For example, in Table 3.3 the elements $1, 5, 7, 11, 13, 17, 19, 23$ of $\mathbb{Z}_{24}^*$ are mapped to the pairs $(1,1), (2,5), (1,7), (2,3), (1,5), (2,1), (1,3), (2,7)$ of $\mathbb{Z}_3^* \times \mathbb{Z}_8^*$.

Lemma 3.4.4. *Assume* $n = n_1 n_2$ *for relatively prime factors* n_1 *and* n_2, *and* $\Phi(a) = (a_1, a_2)$. *Then* $a \in \mathbb{Z}_n^*$ *if and only if* $a_1 \in \mathbb{Z}_{n_1}^*$ *and* $a_2 \in \mathbb{Z}_{n_2}^*$.

Proof. For both directions, we intensively use Proposition 3.1.13.
"$\Rightarrow$": Assume $a \in \mathbb{Z}_n^*$. Write $1 = au + nv$ for integers u and v. We know that $a = a_1 + kn_1$ for $k = a$ div n_1. Hence

$$1 = (a_1 + kn_1)u + nv = a_1 u + (ku + n_2 v)n_1,$$

which implies that $a_1 \in \mathbb{Z}_{n_1}^*$. The proof that $a_2 \in \mathbb{Z}_{n_2}^*$ is identical.
"$\Leftarrow$": Assume $a_1 \in \mathbb{Z}_{n_1}^*$ and $a_2 \in \mathbb{Z}_{n_2}^*$. Write

$$1 = a_1 u_1 + n_1 v_1 \quad \text{and} \quad 1 = a_2 u_2 + n_2 v_2,$$

for suitable integers u_1, v_1, u_2, v_2. By Corollary 3.4.2 there is some $u \in \mathbb{Z}_n^*$ with $u \equiv u_1 \pmod{n_1}$ and $u \equiv u_2 \pmod{n_2}$. Then

$$au \equiv a_1 u_1 \equiv 1 \pmod{n_1} \quad \text{and} \quad au \equiv a_2 u_2 \equiv 1 \pmod{n_2}.$$

Using Corollary 3.4.2 again we conclude that $au \equiv 1 \pmod{n}$, or $a \cdot_n u = 1$, and hence that $a \in \mathbb{Z}_n^*$ by Proposition 3.3.8(c).   □

Corollary 3.4.5. *If* $n = n_1 n_2$ *for* n_1, n_2 *relatively prime, then* $\varphi(n) = \varphi(n_1) \cdot \varphi(n_2)$.

Proof. Using the previous lemma, we have

$$\varphi(n) = |\mathbb{Z}_n^*| = |\mathbb{Z}_{n_1}^*| \cdot |\mathbb{Z}_{n_2}^*| = \varphi(n_1) \cdot \varphi(n_2).$$ □

3.5 Prime Numbers

In this section, we consider prime numbers and establish some basic facts about them. In particular, we state and prove the fundamental theorem of arithmetic, which says that every positive integer can be written as a product of prime numbers in a unique way.

3.5.1 Basic Observations and the Sieve of Eratosthenes

Definition 3.5.1. *A positive integer n is a **prime number** or a **prime** for short if $n > 1$ and there is no number that divides n excepting 1 and n. If n is divisible by some a, $1 < a < n$, then n is a **composite number**.*

Here is a list of the 25 prime numbers between 1 and 100:

$2, 3, 5, 7, 11, 13, 17, 19, 23, 29, 31, 37, 41, 43, 47, 53, 59, 61, 67, 71, 73, 79, 83, 89, 97.$

For example, 2 is a prime number, and 3, but 4 is not, since $4 = 2 \cdot 2$, and 6 is not, since $6 = 2 \cdot 3$.

Here is a first basic fact about the relationship between natural numbers and prime numbers, which is more or less immediate.

Lemma 3.5.2. *If $n \geq 2$, then n is divisible by some prime number.*

Proof. If n has no proper divisor, then n is a prime number, and of course n divides n. Otherwise n has a proper divisor n_1, $1 < n_1 < n$. If n_1 has no proper divisor, then n_1 is a prime number, and n_1 divides n. Otherwise we continue in the same way to see that there is a proper divisor n_2 of n_1, and so on. We obtain a sequence $n > n_1 > n_2 > \cdots$ of divisors of n. Such a sequence cannot be infinite, hence after a finite number of steps we reach a divisor n_t of n with $n_t \geq 2$ that has no proper divisors, hence is a prime number.   □

For example, consider the number $n = 1000$. Our procedure could discover the divisors $500, 250, 50, 10$, in this order, and finally reach the prime factor 2 of 1000. The reader is invited to check that the number of rounds in the procedure just described cannot be larger than $\lfloor \log n \rfloor$. Note however that this does not mean at all that a prime factor of n can always be found in $\log n$ steps by an efficient algorithm. The problem is that it is not known how to find a proper divisor of n fast even if n is known to be composite.

The venerable theorem noted next was proved by Euclid.

Theorem 3.5.3. *There are infinitely many prime numbers.*

Proof. Let $p_1, \ldots, p_s$ be an arbitrary finite list of distinct prime numbers. We form the number $n = p_1 \cdots p_s + 1$. By Lemma 3.5.2 n is divisible by some prime number p. This p must be different from $p_1, \ldots, p_s$. (If p were equal to p_j, then p would be a common divisor of n and $p_1 \cdots p_s = n - 1$, hence p would divide $n - (n - 1) = 1$, which is impossible.) Thus, no finite list of prime numbers can contain all prime numbers.   □

In order to make a list of the prime numbers up to some number n, the easiest way is to use the "Sieve of Eratosthenes". This is an ancient algorithm that can be described informally as follows: Set up a list of all numbers in $\{2, \ldots, n\}$. Initially all numbers are unmarked; some of the numbers will become marked in the course of the calculation. For $j = 2, 3, \ldots, \lfloor \sqrt{n} \rfloor$, do the following: if j is unmarked, then mark all multiples $i = sj$ of j for $j \leq s \leq n/j$.

It is not very hard to see that the numbers that are left unmarked are exactly the prime numbers in the interval $[2, n]$. Indeed, no prime number is ever marked, since only numbers sj with $s \geq j \geq 2$ are marked. Conversely, if $k \leq n$ is composite, then write k as a product ab with $b \geq a \geq 2$. Obviously, then, $a \leq \sqrt{k} \leq \sqrt{n}$. By Lemma 3.5.2 there is some prime number p that divides a. Then we may write $k = sp$ for some $s \geq b$, hence $s \geq p$. When j attains the value p, this prime number turns out to be unmarked, and the number k becomes marked as the multiple sp of p.

A slightly more elaborate version of the Sieve of Eratosthenes, given next, even marks each composite number in $[2, n]$ with its smallest prime divisor.

Algorithm 3.5.4 (The Sieve of Eratosthenes)

INPUT: Integer $n \geq 2$

METHOD:
```
1     m[2..n]: array of integer;
2     for j from 2 to n do m[j] ← 0;
3     j ← 2;
4     while j · j ≤ n do
5         if m[j] = 0 then
6             i ← j · j;
7             while i ≤ n do
8                 if m[i] = 0 then m[i] ← j;
9                 i ← i + j;
10        j ← j + 1;
11    return m[2..n];
```

The algorithm varies the idea sketched above as follows. Instead of attaching a "mark" to k, $m[k]$ is assigned a nonzero value. The j-loop (lines 4–10) treats the numbers $j = 2, 3, \ldots, \lfloor \sqrt{n} \rfloor$, in order. If $m[j]$ turns out to be nonzero when j has value j, then in the i-loop (lines 7–9) the *unmarked* multiples of j are marked with j. — The algorithm is easily analyzed. As before, if $k \leq n$ is a prime number, then the value $m[k]$ stays 0 throughout. Assume now that $k \leq n$ is composite. Consider the smallest prime number p that is a divisor of k, and write $k = sp$ for some $s \geq 2$. By Lemma 3.5.2, s is divisible by some prime number p', which then must exceed p; hence $p^2 \leq sp = k \leq n$. From the algorithm it is then obvious that in the iteration of the j-loop (lines 4–10) in which the variable j contains p the array component $m[k]$ will be assigned the value p. (It cannot get a value different from 0 before that, since this would imply that k is divisible by some number $j < p$, which is impossible by the choice of p.)

What is the time complexity of Algorithm 3.5.4? The number of iterations of the j-loop is bounded by $\sqrt{n}$. Iterations in which j contains a composite number take constant time. Thus we should bound the number of marking steps in the various runs of the i-loop, including those in which the algorithm tries to mark a number already marked. For each prime $p \leq \sqrt{n}$ there are no more than n/p many multiples $sp \leq n$. Hence the total number of marking steps is bounded by

$$\sum_{p \leq \sqrt{n}} \frac{n}{p} \leq n \cdot \sum_{p \leq \sqrt{n}} \frac{1}{p}, \tag{3.5.9}$$

where the sums extend over the prime numbers $p \leq \sqrt{n}$. The last sum is certainly not larger than $\sum_{k \leq \sqrt{n}} \frac{1}{k} < 1 + \ln \sqrt{n} = 1 + \frac{1}{2} \ln n$ (see Lemma A.2.4 in the appendix), hence the number of marking steps is bounded by $O(n \log n)$. (Actually, it is well known that $\sum_{p \leq x} \frac{1}{p} \leq \ln \ln x + O(1)$, for $x \to \infty$, hence the number of marking steps in this naive implementation of the Sieve of Eratosthenes is really $\Theta(n \log \log n)$. See, for example, [31].)

Example 3.5.5. Let $n = 300$. The first few values taken by j are 2 (which leads to all even numbers ≥ 4 being marked with 2), then 3 (which leads to all odd numbers divisible by 3 being marked with 3), then 4, which is marked (with 2), then 5 (which leads to all odd numbers divisible by 5 but not by 3 being marked with 5). This identifies 2, 3, and 5 as the first prime numbers.

	7	**11**	**13**	**17**	**19**	**23**	**29**
31	**37**	**41**	**43**	**47**	7|49	**53**	**59**
61	**67**	**71**	**73**	7|77	**79**	**83**	**89**
7|91	**97**	**101**	**103**	**107**	**109**	**113**	7|119
11|121	**127**	**131**	7|133	**137**	**139**	11|143	**149**
151	**157**	7|161	**163**	**167**	13|169	**173**	**179**
181	11|187	**191**	**193**	**197**	**199**	7|203	11|209
211	7|217	13|221	**223**	**227**	**229**	**233**	**239**
241	13|247	**251**	11|253	**257**	7|259	**263**	**269**
271	**277**	**281**	**283**	7|287	17|289	**293**	13|299

Table 3.4. Result of Sieve of Eratosthenes, $n = 300$

Still unmarked are those numbers larger than 5 that leave remainder 1, 7, 11, 13, 17, 19, 23, or 29 when divided by 30.

In the next round the following multiples of 7 are marked with 7:

$$49, 77, 91, 119, 133, 161, 203, 217, 259, 287.$$

Then the following multiples of 11 (with 11):

$$121, 143, 187, 209, 253.$$

Then the multiples 169, 221, 247, and 299 of 13, with 13, and finally the multiple 289 of 17, with 17. This creates the list in Table 3.4, with the 59 primes in $\{7, 8, \ldots, 300\}$ in bold and composite numbers marked by their smallest prime factor. Note that for the numbers a that do not occur in the list, i.e., the multiples of 2, 3, and 5, it is easy to obtain the smallest prime dividing a from the decimal representation of a.

3.5.2 The Fundamental Theorem of Arithmetic

Once the Sieve of Eratosthenes has been run on an input n, with result $\mathtt{m}[2..n]$, we may rapidly split an arbitrary given number $k \leq n$ into factors that are prime. Indeed, read $i = \mathtt{m}[k]$. If $i = 0$, then k is prime, and we are done. Otherwise $i = p_1$ for p_1 the smallest prime divisor of k. Let $k_1 = k/p_1$ (which is $\leq k/2$). By iterating, find a way of writing $k_1 = p_2 \cdots p_r$ as a product of prime numbers. Then $k = p_1 \cdot p_2 \cdots p_r$ is the desired representation as a product of prime numbers. (In practice, this method is applicable only for k that are so small that we can afford the $O(n \log \log n)$ cost of running the Sieve of Eratosthenes on some $n \geq k$.) A representation $n = p_1 \cdots p_r$ of n as a product of $r \geq 0$ prime numbers $p_1, \ldots, p_r$, not necessarily distinct, is called a **prime decomposition** of n. The number 1 can be represented by the "empty product" of zero factors, a prime number p is represented as a "product" of one factor. Theorem 3.5.8 to follow is the very basic fact about the relationship between the natural numbers and the prime numbers, known and believed since ancient times, proved rigorously by Gauss. It says that every positive integer has one and only one prime decomposition.

Because this result is basic for everything that follows, we give a proof. We start with a lemma that says that a prime decomposition exists. Although we have just seen that this can be deduced by extending the Sieve of Eratosthenes method, we give a short, abstract proof.

Lemma 3.5.6. *Every integer $n \geq 1$ has a prime decomposition.*

Proof. This is proved by induction on n. If $n = 1$, then n is written as the empty product of primes. Thus, assume $n \geq 2$. By Lemma 3.5.2, n can be written $n = p \cdot n'$ for some prime p and some number n'. If $n' = 1$, then $n = p$, and this is the desired prime decomposition. If $1 < n' = n/p < n$, we apply the induction hypothesis to n' to see that there is a prime decomposition $n' = p_1 \cdots p_r$ of n'. Then $n = p \cdot p_1 \cdots p_r$ is the desired prime decomposition of n. □

Lemma 3.5.7. *Let $n \geq 1$, and let p be a prime number. Then p is a divisor of n if and only if n has a prime decomposition in which p occurs as a factor.*

Proof. "$\Leftarrow$": If $n = p_1 \cdots p_r$ with $p = p_j$, then obviously p divides n.
"$\Rightarrow$": By assumption, we may write $n = p \cdot n'$ for some $n' < n$. By Lemma 3.5.6, we may write $n' = p_1 \cdots p_r$ as the product of r prime numbers,

$r \geq 0$. Clearly, then $n = p \cdot p_1 \cdots p_r$ is the desired prime decomposition of n in which p occurs. □

Theorem 3.5.8 (The Fundamental Theorem of Arithmetic).
Every integer $n \geq 1$ can be written as a product of prime numbers in exactly one way (if the order of the factors is disregarded).

Proof. (a) The *existence* of the prime decomposition is given by Lemma 3.5.6.
(b) For the *uniqueness* of the decomposition we argue indirectly. Assume for a contradiction that there is an integer $n \geq 1$ that possesses two different prime decompositions:

$$n = p_1 \cdots p_r = q_1 \cdots q_s, \ r, s \geq 0 .$$

It is clear that $r = 0$ or $s = 0$ is impossible, since then n would have to be 1, and the number 1 has only one prime decomposition (the empty product). We choose an $n \geq 2$ that is minimal with this property. Then we observe that $\{p_1, \ldots, p_r\}$ and $\{q_1, \ldots, q_s\}$ must be disjoint. (If $p_i = q_j$, then n/p_i would have two different prime decompositions, in contradiction to our choosing n minimal with this property.) We may assume that $p_1 < q_1$. (Otherwise interchange the two decompositions.) Now consider the number

$$m = n - p_1 q_2 \cdots q_s = (q_1 - p_1) q_2 \cdots q_s.$$

We have $0 < m < n$. The number m has a prime decomposition without the prime number p_1. (Indeed, $q_1 - p_1$ cannot be divisible by p_1, since otherwise the prime number $q_1 = p_1 + (q_1 - p_1)$ would be divisible by $p_1 < q_1$. By Lemma 3.5.6, $q_1 - p_1$ has a prime decomposition $p_1' \cdots p_t'$, which does not contain p_1. Then the prime decomposition $p_1' \cdots p_t' \cdot q_2 \cdots q_s$ of m does not contain p_1 either.) On the other hand, m is divisible by p_1 (since n and $p_1 q_2 \cdots q_s$ are), which by Lemma 3.5.7 implies that m has a prime decomposition in which p_1 *does* occur. Thus m has two different prime decompositions, contradicting our choosing n minimal with this property. This means that there can be no n with two different prime decompositions. □
    Very often, the prime decomposition of a number n is written as

$$n = p_1^{k_1} \cdots p_r^{k_r}, \tag{3.5.10}$$

where $p_1, \ldots, p_r$, $r \geq 0$, are the *distinct* prime numbers that occur in the prime decomposition of n, and $k_i \geq 1$ is the number of times p_i occurs as a factor in this prime decomposition, for $1 \leq i \leq r$.
    The fundamental theorem entails that the method sketched above for obtaining a prime decomposition on the basis of applying the Sieve of Eratosthenes to n in fact yields the unique prime factorization for $k \leq n$.

Corollary 3.5.9. *If a prime number p divides $n \cdot m$, then p divides n or p divides m.*

Proof. Choose prime decompositions $p_1 \cdots p_r$ of n and $q_1 \cdots q_s$ of m. Then $p_1 \cdots p_r \cdot q_1 \cdots q_s$ is a prime decomposition of $n \cdot m$. By Lemma 3.5.7 and the fundamental theorem (Theorem 3.5.8) p appears among $p_1, \ldots, p_r, q_1, \ldots, q_r$. If p is one of the p_i, then p divides n, otherwise it is among the q_j's and hence divides m. $\qquad\square$

Corollary 3.5.10. *If $p_1, \ldots, p_r$ are distinct prime numbers that all divide n, then $p_1 \cdots p_r$ divides n.*

Proof. Let $n = q_1 \cdots q_s$ be the prime decomposition of n. For each p_i we have the following: By Lemma 3.5.7 and Theorem 3.5.8, p_i must occur among $q_1, \ldots, q_s$. Now the p_i are distinct, so by reordering we may assume $p_i = q_i$, for $1 \leq i \leq r$. Then $n = (p_1 \cdots p_r) \cdot (q_{r+1} \cdots q_s)$, which proves the claim. $\quad\square$

We close this section with a remark on the connection between the concept of numbers being relatively prime and their prime factorization and draw a consequence for the problem of calculating $\varphi(n)$.

Proposition 3.5.11. *Let $n, m \geq 1$ have prime decompositions $n = p_1 \cdots p_r$ and $m = q_1 \cdots q_s$. Then a, b are relatively prime if and only if $\{p_1, \ldots, p_r\} \cap \{q_1, \ldots, q_s\} = \emptyset$.*

Proof. "$\Rightarrow$": Indirectly. Suppose a prime number p occurred in both prime decompositions. Then p would divide both n and m, hence $\gcd(n, m)$ would be larger than 1. — "$\Leftarrow$": Indirectly. Suppose $\gcd(n, m) > 1$. Then there is a prime number p that divides $\gcd(n, m)$ and hence divides both n and m. By Theorem 3.5.8 p occurs in $\{p_1, \ldots, p_r\}$ and in $\{q_1, \ldots, q_s\}$. $\qquad\square$

Corollary 3.4.5 stated that $\varphi(n_1 \cdot n_2) = \varphi(n_1) \cdot \varphi(n_2)$ if n_1 and n_2 are relatively prime. This makes it possible to calculate $\varphi(n)$ for n easily once the prime decomposition of n is known.

Proposition 3.5.12. *If $n \geq 1$ has the prime decomposition $n = p_1^{k_1} \cdots p_r^{k_r}$ for distinct prime numbers $p_1, \ldots, p_r$, then*

$$\varphi(n) = \prod_{1 \leq i \leq r} (p_i - 1) p_i^{k_i - 1} = n \cdot \prod_{1 \leq i \leq r} \left(1 - \frac{1}{p_i} \right).$$

Proof. The formulas are trivially correct for $n = 1$. Thus, assume $n \geq 2$.
Case 1: n is a prime number. — Then $\varphi(n) = |\{1, \ldots, n-1\}| = n - 1$, and the formulas are correct.
Case 2: $n = p^k$ for a prime number p and some $k \geq 2$. — Then

$$\varphi(n) = |\{a \mid 1 \leq a < p^k, p \nmid a\}| = p^k - |\{a \mid 1 \leq a < p^k, p \mid a\}|$$

$$= p^k - p^{k-1} = (p - 1) \cdot p^{k-1} = p^k \cdot \left(1 - \frac{1}{p} \right).$$

Case 3: $n = p_1^{k_1} \cdots p_r^{k_r}$ for $r \geq 2$. — The factors $p_1^{k_1}, \ldots, p_r^{k_r}$ are pairwise relatively prime, by Proposition 3.5.11. We apply Corollary 3.4.5 repeatedly to conclude that

$$\varphi(n) = \prod_{1 \leq i \leq r} \varphi(p_i^{k_i}),$$

which by Case 2 means

$$\varphi(n) = \prod_{1 \leq i \leq r} (p_i - 1) \cdot p^{k_i - 1} = \prod_{1 \leq i \leq r} p^{k_i} \cdot \left(1 - \frac{1}{p_i}\right) = n \cdot \prod_{1 \leq i \leq r} \left(1 - \frac{1}{p_i}\right).$$

$\square$

For example, we have

$$\varphi(7) = 6;$$
$$\varphi(15) = \varphi(3 \cdot 5) = 2 \cdot 4 = 8 = 15 \cdot \tfrac{2}{3} \cdot \tfrac{4}{5},$$
$$\varphi(210) = \varphi(2 \cdot 3 \cdot 5 \cdot 7) = 1 \cdot 2 \cdot 4 \cdot 6 = 48 = 210 \cdot \tfrac{1}{2} \cdot \tfrac{2}{3} \cdot \tfrac{4}{5} \cdot \tfrac{6}{7},$$
$$\varphi(1000) = \varphi(2^3 \cdot 5^3) = 1 \cdot 2^2 \cdot 4 \cdot 5^2 = 400 = 1000 \cdot \tfrac{1}{2} \cdot \tfrac{4}{5},$$
$$\varphi(6860) = \varphi(2^2 \cdot 5 \cdot 7^3) = 1 \cdot 2 \cdot 4 \cdot 6 \cdot 7^2 = 2352 = 6860 \cdot \tfrac{1}{2} \cdot \tfrac{4}{5} \cdot \tfrac{6}{7}.$$

3.6 Chebychev's Theorem on the Density of Prime Numbers

In this section, we develop upper and lower bounds on the density of the prime numbers in the natural numbers. Up to now, we only know that there are infinitely many prime numbers. We want to estimate how many primes there are up to some bound x. Results of the type given here were first proved by Chebychev in 1852, and so they are known as "Chebychev-type estimates".

Definition 3.6.1. *For $x > 1$, let $\pi(x)$ denote the number of primes $p \leq x$.*

Table 3.5 lists some values of this function, and compares it with the function $x/(\ln x - 1)$.

x	2	3	4	5	6	7	8	9	10	11	12	13	14	15	20	30
$\pi(x)$	1	2	2	3	3	4	4	4	4	5	5	6	6	6	8	10

x	40	50	70	100	200	500	1000	5000	10000
$\pi(x)$	12	15	19	25	46	95	168	669	1229
$x/(\ln x - 1)$	14.9	17.2	21.5	27.7	46.5	95.9	169.3	665.1	1218.0

Table 3.5. The prime counting function $\pi(x)$ for some integral values of x; values of the function $x/(\ln x - 1)$ (rounded) in comparison

The following theorem is important, deep, and famous; it was conjectured as early as in the 18th century by Gauss, but proved only in 1896, independently by Hadamard and de la Vallée Poussin.

Theorem 3.6.2 (The Prime Number Theorem).

$$\lim_{x \to \infty} \frac{\pi(x)}{x/\ln x} = 1$$

The prime number theorem should be read as follows: asymptotically, that means for large enough x, about a fraction of 1 in $\ln x$ of the numbers $\leq x$ will be primes, or, the density of prime numbers among the integers in the neighborhood of x is around 1 in $\ln x$. Actually, the figure $x/(\ln x - 1)$ is an even better approximation. We can thus estimate that the percentage of primes in numbers that can be written with up to 50 decimal digits is about $1/\ln(10^{50} - 1) = 1/(50 \ln 10 - 1) \approx 1/114$ or 0.88 percent; for 100 decimal digits the percentage is about $1/\ln(10^{100} - 1) = 1/(100 \ln 10 - 1) \approx 1/229$. In general, doubling the number of digits will approximately halve the percentage of prime numbers. Readers who wish to see a full proof of the prime number theorem are referred to [6]; for details on the quality of the approximation see [16].

We cannot prove the prime number theorem here, and really we do not need it. Rather, we are content with showing that $\pi(x) = \Theta(x/\log x)$, which is sufficient for our purposes. The proofs for these weaker upper and lower bounds are both classical gems and quite clear and should give the reader a good intuitive understanding of why the density of prime numbers in $\{1, \ldots, N\}$ is $\Theta(1/\log N)$. We will have the opportunity to use a variant of these bounds (Proposition 3.6.9) in the analysis of the deterministic primality test. Moreover, lower bounds on the density of the prime numbers are important for analyzing the running time of randomized procedures for generating large prime numbers.

Theorem 3.6.3. *For all integers $N \geq 2$ we have*

$$\frac{N}{\log N} - 2 \leq \pi(N) \leq \frac{3N}{\log N} \ .$$

We prove the lower bound first, and then turn to the upper bound.

Proof of Theorem 3.6.3 — The Lower Bound. First, we focus on even numbers N, of the form $N = 2n$. In the center of the lower bound proof stands the binomial coefficient

$$\binom{2n}{n} = \frac{(2n)!}{n! \cdot n!} = \frac{2n(2n - 1) \cdots (n + 1)}{n(n - 1) \cdots 1} \ .$$

(For a discussion of factorials $n!$ and binomial coefficients $\binom{n}{k}$ see Appendix A.1.) Recall that $\binom{2n}{n}$ is the number of n-element subsets of a $2n$-element set and as such is a natural number. In comparison to $2n$ the number $\binom{2n}{n}$ is very large, namely very close to 2^{2n}. Now consider the prime decomposition

$$\binom{2n}{n} = p_1^{k_1} \cdots p_r^{k_r} .$$

The crucial observation we will make is that for no p_s can the factor $p_s^{k_s}$ in this product be larger than $2n$. To get the big number $\binom{2n}{n}$ as a product of such small contributions requires that the prime decomposition of $\binom{2n}{n}$ contains many different primes — namely, $\Omega(2n/\log(2n))$ many — all of them $\leq 2n$, of course. To extend the estimate to odd numbers $2n + 1$ is only a technical matter.

Next, we will fill in the details of this sketch.

First, we orient ourselves about the order of magnitude of $\binom{2n}{n}$. We have

$$\frac{2^{2n}}{2n} \leq \binom{2n}{n} < 2^{2n} , \text{ for all } n \geq 1. \tag{3.6.11}$$

Roughly, this is because $\sum_{0 \leq i \leq 2n} \binom{2n}{i} = 2^{2n}$ by the binomial theorem, and because $\binom{2n}{n}$ is the largest term in the sum. (For the details, see Lemma A.1.2(c) in Appendix A.1.)

For a number m and a prime p we denote the exact power to which p appears in the prime factorization of m by $\nu_p(m)$. Thus $\nu_p(m)$ is the largest $k \geq 0$ so that $p^k \mid m$, and

$$m = \prod_{p \mid m} p^{\nu_p(m)},$$

where the product extends over all prime factors of m.

For example, $\nu_3(18) = \nu_3(2 \cdot 3^2) = 2$, $\nu_2(10^k) = \nu_5(10^k) = k$. Interestingly, it is almost trivial to calculate to which power a prime divides the number $n!$. This is most easily expressed using the "floor function" (defined in Appendix A.2). To give some intuitive sense to the following formula, note that for integers $a \geq 0$ and $b \geq 1$ the number $\lfloor \frac{a}{b} \rfloor = a \text{ div } b$ equals the number of multiples $b, 2b, 3b, \ldots$ of b that do not exceed a.

Lemma 3.6.4 (Legendre). *For all $n \geq 1$ and all primes p we have*

$$\nu_p(n!) = \sum_{k \geq 1} \left\lfloor \frac{n}{p^k} \right\rfloor .$$

Proof. The proof is a typical example for a simple, but very helpful counting technique used a lot in combinatorics as well as in the amortized analysis of algorithms. Consider the set

$$R_{p,n} = \{(i, k) \mid 1 \leq i \leq n \text{ and } p^k \text{ divides } i \}.$$

Table 3.6 depicts an example for this set ($p = 2$ and $n = 20$) as a matrix with $\log_p(n)$ rows and n columns, and entries 1 ($\bullet$) and 0 (empty). We obtain

	$i=1$	2	3	4	5	6	7	8	9	10	11	12	13	14	15	16	17	18	19	20
$k=1$		•		•		•		•		•		•		•		•		•		•
2				•				•				•				•				•
3								•								•				
4																•				

Table 3.6. The relation $R_{p,n}$ for $p=2$ and $n=20$

two formulas for $|R_{p,n}|$ (corresponding to the number of •'s in the table): counting column-wise, we get

$$|R_{p,n}| = \sum_{1\leq i\leq n} |\{k\geq 1 \mid p^k \text{ divides } i\}| = \sum_{1\leq i\leq n} \max\{k\geq 1 \mid p^k \text{ divides } i\}$$

$$= \sum_{1\leq i\leq n} \nu_p(i) = \nu_p(n!)\,,$$

counting row-wise, we get

$$|R_{p,n}| = \sum_{k\geq 1} |\{i \mid 1\leq i\leq n \text{ and } p^k \text{ divides } i\}| = \sum_{k\geq 1}\left\lfloor\frac{n}{p^k}\right\rfloor\,.$$

Since both results must be the same, the lemma is proved.   □

For example, if $n=20$ and $p=2$, we have $\nu_2(20) = 10+5+2+1 = 18$. Incidentally, the example suggests (and it is easily verified) that the sequence

$$\left\lfloor\frac{n}{p^k}\right\rfloor,\ k=1,2,3,\ldots$$

may be calculated by taking $n_0 = n$, and iteratively dividing by p: $n_1 = n_0 \text{ div } p$, $n_2 = n_1 \text{ div } p$, and so on, until the quotient becomes 0. $\nu_p(n!)$ is then obtained by adding the nonzero n_k, $k\geq 1$.

Lemma 3.6.5. *If p is prime and $\ell = \nu_p(\binom{2n}{n})$, then $p^\ell \leq 2n$.*

Proof. We can calculate the exponent $\ell = \nu_p(\binom{2n}{n})$ of p in $\binom{2n}{n}$ as follows, using Lemma 3.6.4:

$$\ell = \nu_p\left(\binom{2n}{n}\right) = \nu_p\left(\frac{(2n)!}{n!\cdot n!}\right) = \nu_p((2n)!) - 2\nu_p(n!)$$

$$= \sum_{k\geq 1}\left\lfloor\frac{2n}{p^k}\right\rfloor - 2\cdot\sum_{k\geq 1}\left\lfloor\frac{n}{p^k}\right\rfloor = \sum_{k\geq 1}\left(\left\lfloor\frac{2n}{p^k}\right\rfloor - 2\left\lfloor\frac{n}{p^k}\right\rfloor\right).$$

Obviously, in the last sum the summands for k with $p^k > 2n$ are all 0. By Lemma A.2.2, we have $\lfloor 2y\rfloor - 2\lfloor y\rfloor \in \{0,1\}$ for all real numbers $y\geq 0$; hence the summands for k with $1\leq p^k \leq 2n$ are either 0 or 1. Hence

$$\ell \leq \max\{k\geq 1 \mid p^k \leq 2n\}\,, \text{ and } p^\ell \leq 2n. \tag{3.6.12}$$

□

Now we can establish a first connection between the prime counting function $\pi(x)$ and the binomial coefficients.

Lemma 3.6.6. *For all $n \geq 1$,*

$$\binom{2n}{n} \leq (2n)^{\pi(2n)}.$$

Proof. Consider the prime factorization

$$\binom{2n}{n} = p_1^{k_1} \cdots p_r^{k_r}$$

of $\binom{2n}{n}$. The primes that occur in this factorization are factors of $(2n)!$, hence cannot be larger than $2n$. Thus r, the number of different primes occurring in $\binom{2n}{n}$, is not larger than $\pi(2n)$. From the previous lemma we get that each factor $p_s^{k_s}$ is bounded by $2n$. Thus,

$$p_1^{k_1} \cdots p_r^{k_r} \leq (2n)^r \leq (2n)^{\pi(2n)},$$

which proves the lemma.   □

The last lemma enables us to finish the proof of the lower bound in Theorem 3.6.3. Assume first that $N = 2n$ is even. We combine Lemma 3.6.6 with the lower bound $\binom{2n}{n} \geq 2^{2n}/2n$ from inequality (3.6.11) to get

$$(2n)^{\pi(2n)} \geq \frac{2^{2n}}{2n};$$

by taking logarithms and dividing by $\log(2n)$ we obtain $\pi(2n) \geq 2n/\log(2n) - 1$. If $N = 2n + 1$ is odd, we estimate, using the result for $2n$:

$$\pi(2n+1) \geq \pi(2n) \geq \frac{2n}{\log(2n)} - 1 > \frac{2n}{\log(2n+1)} - 1 > \frac{2n+1}{\log(2n+1)} - 2.$$

In either case, we have

$$\pi(N) \geq \frac{N}{\log N} - 2,$$

as claimed.   □

Proof of Theorem 3.6.3 — The Upper Bound. We start with a lemma (first proved by P. Erdös), which states a rough, but very useful upper bound for the products of initial segments of the sequence of prime numbers. As examples note:

$$\begin{aligned}
2 \cdot 3 &= & 6 &< & 16 &= 4^2 \\
2 \cdot 3 \cdot 5 &= & 30 &< & 256 &= 4^4 \\
2 \cdot 3 \cdot 5 \cdot 7 &= & 210 &< & 4096 &= 4^6 \\
2 \cdot 3 \cdot 5 \cdot 7 \cdot 11 &= & 3210 &< & 1048576 &= 4^{10}
\end{aligned}$$

In contrast, note that for constant α, $0 < \alpha < 1$, and $c > 1$ the product $N!$ of all integers from 1 up to αN becomes (much) larger than c^N for sufficiently large N, no matter how small α and how large c are chosen. To see this, consider that inequality (A.1.2) says $(\alpha N)! > (\alpha N/e)^{\alpha N}$, hence $(\alpha N)!/c^N > ((N/C)^\alpha)^N$, for a suitable constant C. Thus the following innocent-looking lemma already makes it clear that the density of the prime numbers below N must be significantly smaller than a constant fraction.

Lemma 3.6.7. *If $N \geq 2$, then*

$$\prod_{p \leq N} p < 4^{N-1},$$

where the product extends over all prime numbers $p \leq N$.

Proof. Again, in the center of the proof are cleverly chosen binomial coefficients, this time the numbers

$$b_m = \binom{2m+1}{m} = \frac{(2m+1)!}{(m+1)!m!} = \frac{(2m+1)(2m)\cdots(m+2)}{m!}, \qquad (3.6.13)$$

for $m \geq 1$. (Examples: $b_1 = 3$, $b_2 = 10 = 2 \cdot 5$, $b_3 = 35 = 5 \cdot 7$, $b_4 = 126 = 2 \cdot 3^2 \cdot 7$, $b_5 = 462 = 2 \cdot 3 \cdot 7 \cdot 11$, and so on.) The number b_m is divisible by all prime numbers p, for $m+2 \leq p \leq 2m+1$. Indeed, in the rightmost fraction in (3.6.13) the numerator contains all these prime numbers as explicit factors, the denominator cannot contain any of them. This immediately leads to the upper bound

$$\prod_{m+2 \leq p \leq 2m+1} p \leq b_m, \qquad (3.6.14)$$

where the product extends over the primes p in $\{m+2, \ldots, 2m+1\}$. An upper bound for b_m is obtained as follows: We know (see (A.1.4) in Appendix A.1) that

$$\sum_{1 \leq i \leq 2m} \binom{2m+1}{i} = 2^{2m+1} - 2, \qquad (3.6.15)$$

and observe that $b_m = \binom{2m+1}{m} = \binom{2m+1}{m+1}$ occurs twice in this sum (see Lemma A.1.2(a)). Hence $b_m < \frac{1}{2} \cdot 2^{2m+1} = 4^m$. Combining this with (3.6.14) we obtain

$$\prod_{m+2 \leq p \leq 2m+1} p < 4^m, \text{ for } m \geq 1. \qquad (3.6.16)$$

Now we may prove the claimed inequality $\prod_{p \leq N} p < 4^{N-1}$ for integers $N \geq 2$, by induction on N.

Initial step: For $N = 2$ we observe that $2 < 4^1$.

Induction step: Assume $N \geq 3$ and the claim is true for all $m < N$.

Case 1: N is even. — Then N is not prime, hence

$$\prod_{p \leq N} p = \prod_{p \leq N-1} p < 4^{N-2} < 4^{N-1},$$

by the induction hypothesis applied to $m = N - 1$.

Case 2: N is odd. — Then we write $N = 2m + 1$, and apply the induction hypothesis to $m + 1$ (note that $2 \leq m + 1 < N$), and then (3.6.16) to obtain

$$\prod_{p \leq N} p = \prod_{p \leq m+1} p \cdot \prod_{m+2 \leq p \leq 2m+1} p < 4^m \cdot 4^m = 4^{2m} = 4^{N-1}. \qquad \square$$

Lemma 3.6.8. *Let $p_1, p_2, p_3, \ldots$ be the sequence of prime numbers in ascending order. Then $p_1 \cdots p_k \geq 2^k \cdot k!$, for all $k \geq 9$.*

Proof. We may check by direct calculation that $p_9 = 23$ and that $p_1 \cdots p_9 = 2 \cdot 3 \cdot 5 \cdots 19 \cdot 23 = 223092870 > 185794560 = 2^9 \cdot 9!$. For larger k, we proceed by induction. Assume $k \geq 9$ and the lemma is true for k. Clearly, we have $p_{k+1} > 2(k+1)$. Thus,

$$p_1 \cdots p_{k+1} = p_1 \cdots p_k \cdot p_{k+1} > 2^k \cdot k! \cdot 2(k+1) = 2^{k+1} \cdot (k+1)!,$$

which is the induction step. $\qquad \square$

Now, at last, we are ready to prove the upper bound $\pi(N) < 3N/\log N$ from Theorem 3.6.3. This inequality is easily checked by inspection for $2 \leq N \leq 26$, so we may assume that $N \geq 27$. Let $k = \pi(N)$, and let $p_1, \ldots, p_k$ be the prime numbers not exceeding N. By Lemma 3.6.8, we have

$$\prod_{p \leq N} p = p_1 \cdots p_k > 2^k \cdot k!. \tag{3.6.17}$$

As noted in Appendix A.1 (see inequality (A.1.2)), we have $k! > (k/e)^k$. Combining this with (3.6.17) and Lemma 3.6.7 yields

$$4^N > 2^k \cdot \left(\frac{k}{e}\right)^k, \tag{3.6.18}$$

or, taking logarithms,

$$(2\ln 2) \cdot N > k \cdot (\ln k + \ln 2 - 1). \tag{3.6.19}$$

We use an indirect argument to show that $k < 2N/\ln N$. (Since $3/\log N = 3\ln 2/\ln N > 2.07/\ln N$, this is sufficient.) Thus, assume for a contradiction that $k \geq 2N/\ln N$. Substituting this into (3.6.19) we obtain

$$(2\ln 2) \cdot N > \frac{2N}{\ln N} \cdot (\ln 2 + \ln N - \ln \ln N + \ln 2 - 1),$$

or, by obvious transformations,

$$(1 - \ln 2) \ln N < \ln \ln N - 2 \ln 2 + 1. \tag{3.6.20}$$

Now the function $f \colon x \mapsto (1 - \ln 2) \ln x - \ln \ln x + 2 \ln 2 - 1$, which is defined for $x > 1$, satisfies $f(27) > 0.2$ and $f'(x) = (1 - \ln 2)/x - 1/(x \ln x)$. This derivative has only one root, which is $e^{1/(1 - \ln 2)} \approx 26.02$, at which point it changes from the negative to the positive. So $f(x) > 0$ for all $x \geq 27$, contradicting (3.6.20). Thus the assumption is wrong, and $\pi(N) < 3N/\log N$ must be true. $\qquad\Box$

To close this section on Chebychev-type inequalities, we use results and methods developed so far to prove an exponential lower bound on the product $\prod_{p \leq N} p$, a kind of mirror image of Lemma 3.6.7, which will be needed in the time analysis of the deterministic primality test in Chap. 8.

Proposition 3.6.9. $\prod_{p \leq 2n} p > 2^n$, for all $n \geq 2$, where the product extends over all primes $p \leq 2n$.

Proof. We know that $2^{2n}/(2n) < \binom{2n}{n}$, see Lemma A.1.2(c). In variation of the proof of Lemma 3.6.6, we find an upper bound on this binomial coefficient, as follows. Again, consider the prime factorization $\binom{2n}{n} = p_1^{k_1} \cdots p_r^{k_r}$. If $p_i \leq \sqrt{2n}$, we are satisfied with the estimate $p_i^{k_i} \leq 2n$ from Lemma 3.6.5. If $p_i > \sqrt{2n}$, then $p_i^2 > 2n$, and by (3.6.12) we conclude $k_i = 1$. This implies

$$2^{2n}/(2n) < \binom{2n}{n} \leq \prod_{p \leq \sqrt{2n}} 2n \cdot \prod_{\sqrt{2n} < p \leq 2n} p.$$

Let us abbreviate $\prod_{p \leq 2n} p$ by Π_{2n}. Then the last inequality implies that

$$2^{2n}/(2n) < (2n)^{\pi(\sqrt{2n})} \cdot \Pi_{2n} < (2n)^{3\sqrt{2n}/\log(\sqrt{2n})} \cdot \Pi_{2n},$$

by the upper bound in Theorem 3.6.3. Since $(2n)^{1/\log(\sqrt{2n})} = (2n)^{2/\log(2n)} = 2^2$, the last inequality means that

$$\Pi_{2n} > 2^{2n} / (2n \cdot 2^{6\sqrt{2n}}).$$

To prove Proposition 3.6.9, we must show that the last quotient exceeds 2^n. This means, we must show that

$$2^n \geq 2n \cdot 2^{6\sqrt{2n}}.$$

Taking logarithms, this amounts to

$$n - 1 - \log n - 6\sqrt{2n} \geq 0. \tag{3.6.21}$$

Clearly, for n large enough, this is true. We show that (3.6.21) is true for $n \geq 100$, by calculus. If we let $f(x) = x - 1 - \log x - 6\sqrt{2x}$, then $f(100) =$

n	$2n$	Π_{2n}		
2	4	$2 \cdot 3 = 6$	>	2^2
3	6	$2 \cdot 3 \cdot 5 = 30$	>	2^4
5	10	$2 \cdot 3 \cdot 5 \cdot 7 = 210$	>	2^7
8	16	$2 \cdot 3 \cdot 5 \cdot 7 \cdot 11 \cdot 13 = 30030$	>	2^{14}
15	30	$> 2^{14} \cdot 17 \cdot 19 \cdot 23 \cdot 29 > 2^{14} \cdot (2^4)^4$	=	2^{30}
30	60	$> 2^{30} \cdot 31 \cdot 37 \cdot \ldots \cdot 59 > 2^{30} \cdot (2^5)^7$	=	2^{65}
65	130	$> 2^{65} \cdot 61 \cdot 67 \cdot 71 \cdot \ldots \cdot 113 \cdot 127 > 2^{65} \cdot (2^6)^{14}$	=	2^{149}

Table 3.7. Lower bounds for products of initial segments of the prime numbers

$99 - \log 100 - 6\sqrt{200} > 8$, and the derivative $f'(x) = 1 - \frac{1}{x} - \frac{6}{\sqrt{2x}}$ is larger than $1 - \frac{1}{100} - \frac{6}{10\sqrt{2}} > 0$ for $x \geq 100$.

Finally, for $n < 100$, we use inspection. Table 3.7 gives all the required information. For establishing the table we have used the fact that between 30 and 60 there are 7 prime numbers (and that $31 \cdot 37 > 2^{10}$), between 60 and 130 there are 14 (and that $61 \cdot 71 > 2^{12}$); see the Sieve of Eratosthenes, Table 3.4. □

4. Basics from Algebra: Groups, Rings, and Fields

In this chapter we develop basic algebraic notions and facts to the extent needed for the applications in this book. Equally important are the examples for such structures from number theory. At the center of attention are basic facts from group theory, especially about cyclic groups, which are central in the analysis of the deterministic primality test. We discuss (commutative) rings (with 1), with the central example being $\mathbb{Z}_m$. Finally, we develop the basic facts about finite fields, in particular we establish that in finite fields the multiplicative group is cyclic.

4.1 Groups and Subgroups

If A is a set, a **binary operation** $\circ$ on A is a mapping $\circ : A \times A \to A$. In the context of groups, we use infix notation for binary operations, i.e., we write $a \circ b$ for $\circ(a, b)$. Examples of binary operations are the addition and the multiplication operation on the set of positive integers or on the set $\mathbb{Z}$.

Definition 4.1.1. *A **group** is a set G together with a binary operation $\circ$ on G with the following properties:*

(i) (**Associativity**) $(a \circ b) \circ c = a \circ (b \circ c)$, *for all* $a, b, c \in G$.
(ii) (**Neutral element**) *There is an element $e \in G$ that satisfies $a \circ e = e \circ a = a$ for each $a \in G$. (In particular, G is not empty.)*
(iii) (**Inverse element**) *For each $a \in G$ there is some $b \in G$ such that $a \circ b = b \circ a = e$ (for the neutral element e from (b)).*

In short, we write $(G, \circ, e)$ for a group with these components.

In view of the associative law, we can put parentheses at any place we want in expressions involving elements $a_1, \ldots, a_r$ of G and the operation $\circ$, without changing the element denoted by such an expression. For example, $(a_1 \circ a_2) \circ (a_3 \circ (a_4 \circ a_5)) = a_1 \circ ((a_2 \circ (a_3 \circ a_4)) \circ a_5)$. In consequence, we will usually omit parentheses altogether, and simply write $a_1 \circ a_2 \circ a_3 \circ a_4 \circ a_5$ for this element.

Groups are abundant in mathematics (and in computer science). Here are a few examples.

M. Dietzfelbinger: Primality Testing in Polynomial Time, LNCS 3000, pp. 55-71, 2004.
© Springer-Verlag Berlin Heidelberg 2004

Example 4.1.2. (a) The set $\mathbb{Z}$ with integer addition as operation and 0 as
neutral element is a group.

(b) For each positive integer m, the set $m\mathbb{Z} = \{m \cdot z \mid z \in \mathbb{Z}\}$ of all multiples
of m, with integer addition as operation and 0 as neutral element, is a
group.

(c) For each integer $n > 1$ the set $\mathbb{Z}_n = \{0, 1, \ldots, n - 1\}$ with addition
modulo n as operation and 0 as neutral element is a group. (See Definition 3.3.1.) The set $\{0\}$ with the operation $0 \circ 0 = 0$ also is a group (the
trivial group).

(d) For each integer $n > 1$ the set $\mathbb{Z}_n^* = \{a \mid 0 \le a < n, \gcd(a, n) = 1\}$
with multiplication modulo n and 1 as neutral element is a group. (See
Definition 3.3.7 and Example 3.3.9.)

(e) Let S be an arbitrary set, and consider the set $\mathrm{Bij}(S)$ of all bijective
mappings $f \colon S \to S$. The operation $\circ$ denotes the composition of mappings
(i.e., $f \circ g(x) = f(g(x))$ for all $x \in S$, $f, g \in G$). Then $(\mathrm{Bij}(S), \circ, \mathrm{id}_S)$ forms
a group, with $\mathrm{id}_S : S \ni x \mapsto x \in S$, the identity map, as neutral element.

(f) Let $\mathrm{GL}_n(\mathbb{Q})$ denote the set of all invertible $n \times n$-matrices over the field $\mathbb{Q}$
of rational numbers. Let $\circ_n$ denote the multiplication of such matrices, and
let I_n denote the $n \times n$ identity matrix (1 on all positions of the diagonal,
0 everywhere else). Then $(\mathrm{GL}_n(\mathbb{Q}), \circ_n, I_n)$ is a group.

Notation: In the cases (a), (b), and (c), the group operation is written as
"$+$" or $+_m$, and the neutral element as 0.

For small groups, we may describe the group operation by writing down
or storing a table with rows and columns indexed by the elements of G, the
element in row a and column b being $a \circ b$. For example, the group table of
$(\mathbb{Z}_9^*, \cdot_9, 1)$ looks as follows.

$\cdot_9$	1	2	4	5	7	8
1	1	2	4	5	7	8
2	2	4	8	1	5	7
4	4	8	7	2	1	5
5	5	1	2	7	8	4
7	7	5	1	8	4	2
8	8	7	5	4	2	1

Table 4.1. Operation table of a group. The group operation is $\cdot_9$, multiplication
modulo 9 on the set $\{1, 2, 4, 5, 7, 8\}$

Obviously, for larger groups such an explicit representation is unfeasible; and as soon as the number of elements of the group is a number with
20 decimal digits, not even one line of the group table can be stored in a
computer.

We remark that there are many extremely important groups in mathematics and in application areas that are *not commutative* in the sense that

there are elements $a, b \in G$ with $a \circ b \neq b \circ a$. For example, in Example 4.1.2(d) (set of bijections from S to S) there are such elements as soon as $|S| \geq 3$; in Example 4.1.2(e) (invertible matrices) there are such elements as soon as $n \geq 2$. In this book, though, we will be dealing exclusively with groups in which such a thing does not occur.

Definition 4.1.3. *We say a group $(G, \circ, e)$ is **commutative** or **abelian** if $a \circ b = b \circ a$ for all $a, b \in G$.*

The groups from Example 4.1.2(a), (b), and (c) are abelian. In abelian groups, in expressions involving elements of G and the operation $\circ$, we may change the order arbitrarily without affecting the result.

We list some facts that hold for all groups, commutative or not, and follow easily from the definitions.

Proposition 4.1.4. (a) *In a group, there is exactly one neutral element (called e or 1 from here on).*
(b) *For each element a of a group G, there is exactly one $b \in G$ such that $a \circ b = b \circ a = e$. (This element is denoted a^{-1} from here on.)*
(c) *(**Cancellation rule**) If $a \circ c = b \circ c$, then $a = b$. Likewise, if $c \circ a = c \circ b$, then $a = b$.*

Proof. (a) If e and e' are neutral elements, i.e., satisfy (ii) from Definition 4.1.1, then we get $e' = e' \circ e = e$ by using first that e is neutral and then that e' is neutral.
(b) If b and b' are inverse to a, i.e. satisfy (iii) from Definition 4.1.1, then we get

$$b = b \circ e = b \circ (a \circ b') = (b \circ a) \circ b' = e \circ b' = b',$$

by using in addition that e is neutral and associativity.
(c) Assume $a \circ c = b \circ c$. Then, by associativity,

$$a = a \circ (c \circ c^{-1}) = (a \circ c) \circ c^{-1} = (b \circ c) \circ c^{-1} = b \circ (c \circ c^{-1}) = b.$$

In the other case, we multiply with c^{-1} from the left. $\square$

Let $a \in G$, and consider a^{-1}. Since $a \circ a^{-1} = a^{-1} \circ a = e$, a is the inverse of a^{-1}, in short $(a^{-1})^{-1} = a$.

Notation: In the situation of Example 4.1.2(a), (b), and (c), where we use additive notation for the groups, the inverse of a is denoted by $-a$. Thus, $a + (-a) = (-a) + a = 0$. For $a + (-b)$ we write $a - b$.

Definition 4.1.5. *Let $(G, \circ, e)$ be a group. A set $H \subseteq G$ is called a **subgroup** of G if H together with the operation $\circ$ and the neutral element inherited from $(G, \circ, e)$ forms a group. More exactly, we require that*

(i) *$e \in H$,*
(ii) *$a \circ b \in H$ for all $a, b \in H$,*
(iii) *$a^{-1} \in H$ for all $a \in H$.*

In Example 4.1.2(b), $m\mathbb{Z}$ is a subgroup of $(\mathbb{Z}, +, 0)$, for each positive integer m. In contrast, $m\mathbb{Z}$ is a subgroup of $n\mathbb{Z}$ if and only if $n \mid m$.

Quite often, we will have to prove that some subset H of a *finite* group G is in fact a subgroup. For this, we provide an easy-to-apply criterion.

Lemma 4.1.6. *If $(G, \circ, e)$ is a finite group, and H is a subset of G with*

(i) $e \in H$, *and*
(ii) H *is closed under the group operation $\circ$,*

then H is a subgroup of G.

Note that the condition that G is finite is necessary to draw this conclusion, since, for example, $\mathbb{N}$ is a subset of $\mathbb{Z}$ that contains 0 and is closed under addition, but $\mathbb{N}$ is not a subgroup of $(\mathbb{Z}, +, 0)$.

Proof. We must check condition (iii) of Definition 4.1.5. For an arbitrary element $a \in H$, consider the mapping

$$f_a \colon H \to H, \; b \mapsto a \circ b,$$

which is well defined by (ii). Since G is a group, f_a is one-to-one (indeed, if $f_a(b_1) = f_a(b_2)$, i.e., $a \circ b_1 = a \circ b_2$, then $b_1 = b_2$ by the cancellation rule). Because H is finite, f_a is a bijection of H onto itself. Using (i) it follows that there is an element $c \in H$ with $a \circ c = f_a(c) = e$; this means that $c = a^{-1} \in H$, and condition (iii) in Definition 4.1.5 is established.   □

A subgroup H splits the elements of a group G into disjoint classes.

Definition 4.1.7. *Let H be a subgroup of a group G. Define*

$$a \sim_H b \;, \; \text{if } b^{-1} \circ a \in H.$$

Lemma 4.1.8. (a) *$\sim_H$ is an equivalence relation.*
(b) *For each $b \in G$, there is a bijection between H and the equivalence class $[b]_H$ of b.*

Proof. (a) *Reflexivity:* $a^{-1} \circ a = e \in H$. *Symmetry:* If $b^{-1} \circ a \in H$, then $a^{-1} \circ b = (b^{-1} \circ a)^{-1} \in H$. *Transitivity:* $b^{-1} \circ a \in H$ and $c^{-1} \circ b \in H$ implies $c^{-1} \circ a = (c^{-1} \circ b) \circ (b^{-1} \circ a) \in H$.
(b) Let $[b]_H = \{a \in G \mid a \sim_H b\}$ be the equivalence class of b. Consider the mapping

$$g_b \colon [b]_H \to G, \; a \mapsto b^{-1} \circ a.$$

By the very definition of $[b]_H$ and of $\sim_H$ we have that $g_b(a) \in H$ for all $a \in [b]_H$. We show that actually g_b maps $[b]_H$ one-to-one onto H: Every $c \in H$ occurs in the image $g_b([b]_H)$, since $b \circ c \in [b]_H$ and $g_b(b \circ c) = c$. Further, g_b is one-to-one, since $g_b(a) = g_b(a')$ implies $a = b \circ g_b(a) = b \circ g_b(a') = a'$.   □

Note that the bijection g_b depends on b, but this is not important. We mention two examples. — As noted above, for $m \geq 1$ the group $m\mathbb{Z}$ is a subgroup of $\mathbb{Z}$. Two elements a and b are equivalent if $(-b) + a \in m\mathbb{Z}$, i.e., if m is a divisor of $a - b$. This is the case if and only if $a \equiv b \pmod{m}$. The equivalence classes are just the classes of numbers that are congruent modulo m. The bijection g_b from $[b]$ to $m\mathbb{Z}$ is given by $a \mapsto a - b$. — As a second example, consider the group $\mathbb{Z}_{24}$ with addition modulo 24. Then the set $\{0, 6, 12, 18\}$ forms a subgroup, since it is closed under the group operation. The equivalence class of 11 is $\{5, 11, 17, 23\}$, and the bijection g_{11} is given by

$$5 \mapsto 5 - 11 \bmod 24 = 18, \quad 11 \mapsto 0, \quad 17 \mapsto 6, \quad 23 \mapsto 12.$$

In the case of finite groups, the existence of a bijection between H and $[b]_H$ has an important consequence that will be essential in the analysis of the randomized primality tests.

Proposition 4.1.9. *If H is a subgroup of the finite group G, then $|H|$ divides $|G|$.*

Proof. Let $C_1, \ldots, C_r$ be the distinct equivalence classes w.r.t. $\sim_H$. They partition G, hence $|G| = |C_1| + \cdots + |C_r|$. Clearly, H appears as one of the equivalence classes (namely $[e]_H = H$), so we may assume that $C_1 = H$. By Lemma 4.1.8(b) we have $|C_1| = \cdots = |C_r|$, and conclude $|G| = r \cdot |H|$.   □

4.2 Cyclic Groups

The concept of a cyclic group is omnipresent in the remainder of the book, because it is central in the analysis of the deterministic primality test.

4.2.1 Definitions, Examples, and Basic Facts

We start by considering powers of an element in arbitrary groups. Let $(G, \circ, e)$ be a group. As an abbreviation, we define for every $a \in G$:

$$a^i = \underbrace{a \circ \cdots \circ a}_{i \text{ times}} \quad \text{and} \quad a^{-i} = \underbrace{a^{-1} \circ \cdots \circ a^{-1}}_{i \text{ times}},$$

for $i \geq 0$, or, more formally, by induction,

$$a^0 = e,$$
$$a^i = a \circ a^{i-1}, \text{ for } i \geq 1,$$

and $a^{-i} = (a^{-1})^i$, for $i \geq 1$. It is a matter of routine to establish the usual laws of calculating with exponents.

Lemma 4.2.1. (a) $(a^i)^{-1} = a^{-i}$, for $a \in G$, $i \in \mathbb{Z}$;
(b) $a^{i+j} = a^i \circ a^j$, for $a \in G$, $i, j \in \mathbb{Z}$;
(c) if $a, b \in G$ satisfy $a \circ b = b \circ a$, then $(a \circ b)^i = a^i \circ b^i$, for $i \in \mathbb{Z}$.

Proof. (a) If $i = 0$, there is nothing to show. We first turn to the case $i > 0$. Consider $c = a^i \circ a^{-i} = a \circ \cdots \circ a \circ a^{-1} \circ \cdots \circ a^{-1}$, with a and a^{-1} each repeated i times. We can combine $a \circ a^{-1}$ to obtain e and then omit the factor e to see that $c = a^i \circ a^{-i} = a \circ \cdots \circ a \circ a^{-1} \circ \cdots \circ a^{-1}$, with a and a^{-1} each repeated $i - 1$ times. Iterating this process we obtain $c = e$. This means that a^{-i} is the (unique) inverse of a^i, as claimed. Finally, if $i < 0$, we apply the claim for the positive exponent $-i$ to get $(a^{-i})^{-1} = a^{-(-i)} = a^i$. By Proposition 4.1.4(b) we conclude that a^{-i} is the unique inverse of a^i in this case as well.
(b) If $i = 0$ or $j = 0$, there is nothing to show. If $i, j > 0$ or $i, j < 0$, the definition and the associative law are enough to prove the claim. Thus, assume $i > 0$ and $j < 0$. Let $k = -j$. We must show that $a^i \circ a^{-k} = a^{i-k}$. If $i = k$, this was proved in (a). Now consider the case $i > k$. Then, using associativity and (a), we get

$$a^i \circ a^{-k} = (a^{i-k} \circ a^k) \cdot a^{-k} = a^{i-k} \circ (a^k \circ a^{-k}) = a^{i-k} \circ e = a^{i-k}.$$

Finally, if $i < k$, then

$$a^i \circ a^{-k} = (a^i \circ (a^{-1})^i) \circ (a^{-1})^{k-i} = (a^{-1})^{k-i} = a^{-(k-i)} = a^{i-k}.$$

(c) If $i = 0$, there is nothing to show. If $i > 0$, then

$$(a \circ b)^i = \underbrace{(a \circ b) \circ \cdots \circ (a \circ b)}_{i \text{ times}}.$$

By using associativity and the fact that we can interchange a and b, this can be transformed into $a \circ \cdots \circ a \circ b \circ \cdots \circ b = a^i \circ b^i$. Now we turn to the case of negative exponents. Note first that $(b \circ a) \circ (a^{-1} \circ b^{-1}) = e$, which means that $(a \circ b)^{-1} = (b \circ a)^{-1} = a^{-1} \circ b^{-1}$. By symmetry, $(b \circ a)^{-1} = b^{-1} \circ a^{-1}$, which means that also $a^{-1} \circ b^{-1} = b^{-1} \circ a^{-1}$. Thus, for $i = -k < 0$, we may apply our result for positive exponents to get

$$(a \circ b)^i = ((a \circ b)^{-1})^k = (a^{-1} \circ b^{-1})^k = (a^{-1})^k \circ (b^{-1})^k = a^i \circ b^i,$$

as desired. $\square$

Note that (b) in particular says that $a^i \circ a^j = a^j \circ a^i$ for all integers i and j, so among arbitrary powers of a we have commutativity. Because they are so natural, the rules listed in Lemma 4.2.1 will be used without further comment in what follows.

Proposition 4.2.2. *Let* $(G, \circ, e)$ *be a group. For* $a \in G$ *define*

$$\langle a \rangle = \{a^i \mid i \in \mathbb{Z}\} = \{e, a, a^{-1}, a^2, (a^{-1})^2, a^3, (a^{-1})^3, \dots\}.$$

Then $\langle a \rangle$ *is a (commutative) subgroup of* G *and it contains* a. *In fact, it is the smallest subgroup of* G *with this property. (It is called the subgroup* **generated by** a.)

Proof. Clearly, $\langle a \rangle$ contains $a = a^1$ and $e = a^0$. From the previous lemma it follows that it is a subgroup (with a^i and a^j it contains $a^i \cdot a^j = a^{i+j}$; and with a^i it contains the inverse a^{-i}). If H is any subgroup of G that contains a, then all elements a^i must be in H, hence we have $\langle a \rangle \subseteq H$. □

Definition 4.2.3. *We say a group $(G, \circ, e)$ is **cyclic** if there is an $a \in G$ such that $G = \langle a \rangle$. An element $a \in G$ with this property is called a **generating element** of G.*

Example 4.2.4. (a) $(\mathbb{Z}, +, 0)$ is a cyclic group, with generating elements 1 and -1.

(b) For $m \geq 1$, the additive group $(\mathbb{Z}_m, +_m, 0)$ is a cyclic group, where $+_m$ denotes addition modulo m. Clearly, 1 is a generator, but there are others: Let $i \in \mathbb{Z}_m$ with $\gcd(i, m) = 1$. (We know that there are $\varphi(m)$ such numbers.) Now $0, i, (i+i) \bmod m, (i+i+i) \bmod m, \ldots, (\underbrace{i + \cdots + i}_{m-1 \text{ times}}) \bmod m$

are all different, and hence exhaust $\mathbb{Z}_m$. Indeed, if $ki \bmod m = \ell i \bmod m$, with $0 \leq k < \ell < m$, then $(\ell - k)i \equiv 0 \pmod{m}$. Since $\gcd(i, m) = 1$, we must have that m divides $\ell - k$ and hence that $\ell = k$. On the other hand, if $d = \gcd(i, m) > 1$, then we get $(m/d)i = (m/d) \cdot (qd) = mq \equiv 0 \pmod{m}$, and hence i cannot generate $\mathbb{Z}_m$.

(c) A not so obvious cyclic group is $\mathbb{Z}_9^* = \{1, 2, 4, 5, 7, 8\}$ with multiplication modulo 9 (see Table 4.1). By direct calculation we see that the powers $5^i \bmod 9$, $0 \leq i < 6$, in this order are $1, 5, 7, 8, 4, 2$, and hence that 5 is a generator. This observation makes it clear that the structure of this group is quite simple, which also becomes apparent in the operation table if the elements are arranged in a suitable order, as in Table 4.2.

$\cdot_9$	1	5	7	8	4	2
1	1	5	7	8	4	2
5	5	7	8	4	2	1
7	7	8	4	2	1	5
8	8	4	2	1	5	7
4	4	2	1	5	7	8
2	2	1	5	7	8	4

Table 4.2. A group operation table of a cyclic group. The group operation is multiplication modulo 9 on the set $\{1, 2, 4, 5, 7, 8\}$

Although we do not prove it here, it is a fact that *all* groups $\mathbb{Z}_{p^\ell}^*$, where p is an odd prime number, are cyclic.

(d) The set U_r of "rth roots of unity" in the field $\mathbb{C}$ is the set of all solutions of the equation $x^r = 1$ in $\mathbb{C}$. It is well known that

$$U_r = \{e^{i \cdot s \cdot 2\pi/r} \mid 0 \leq s < r\},$$

where i is the imaginary unit. If $\mathbb{C}$ is depicted as the Euclidean plane, the set U_r appears as an equidistant grid of r points on the unit circle, containing 1. The elements are multiplied according to the rule

$$e^{i \cdot s \cdot 2\pi/r} \cdot e^{i \cdot t \cdot 2\pi/r} = e^{i \cdot (s+t) \cdot 2\pi/r} = e^{i \cdot ((s+t) \bmod r) \cdot 2\pi/r},$$

which corresponds to the addition of angles, ignoring multiples of 2π.

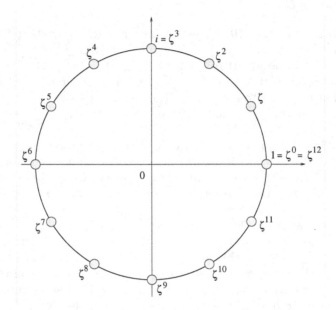

Fig. 4.1. The cyclic group of the rth roots of unity in $\mathbb{C}$, for $r = 12$

With $\zeta = e^{i \cdot 2\pi/r}$ the natural generating element, we have $U_r = \{1, \zeta, \zeta^2, \ldots, \zeta^{r-1}\}$. We shall see later that all cyclic groups of size r are isomorphic; thus, the depiction given in Fig. 4.1 applies for every finite cyclic group.

Clearly, for an arbitrary group G and every $a \in G$ the subgroup $\langle a \rangle$ is cyclic.

Definition 4.2.5. *Let $(G, \circ, e)$ be a group. The **order** $\operatorname{ord}_G(a)$ of an element $a \in G$ is defined as*

$$\begin{cases} |\langle a \rangle|, & \text{if } \langle a \rangle \text{ is finite,} \\ \infty, & \text{otherwise.} \end{cases}$$

4.2.2 Structure of Cyclic Groups

The following proposition shows that in fact there are only two different types of cyclic groups: finite and infinite ones. The infinite ones have the same structure as $(\mathbb{Z}, +, 0)$, the finite ones have the structure of some $(\mathbb{Z}_m, +, 0)$. In this text, only finite groups are relevant.

Lemma 4.2.6. *Let $(G, \circ, e)$ be a group, and let $a \in G$.*

(a) *If all elements a^i, $i \in \mathbb{Z}$, are different, then $\mathrm{ord}_G(a) = \infty$ and the group $\langle a \rangle$ is isomorphic to $\mathbb{Z}$ via the mapping $i \mapsto a^i$, $i \in \mathbb{Z}$.*
(b) *If $a^i = a^j$ for integers $i < j$, then $\mathrm{ord}_G(a)$ is finite and $\mathrm{ord}_G(a) \leq j - i$.*

Proof. (a) Assume that all a^i, $i \in \mathbb{Z}$, are different. Then the mapping $i \mapsto a^i$ is a bijection between $\mathbb{Z}$ and $\langle a \rangle$. That it is also an isomorphism between $(\mathbb{Z}, +, 0)$ and $(\langle a \rangle, \circ, e)$, i.e., that 0 is mapped to e and $i + j$ to $a^i \circ a^j$ and $-i$ to $(a^i)^{-1}$, corresponds to the rules in Lemma 4.2.1(a) and (b).
(b) Assume that $a^i = a^j$ for $i < j$. Then for $k = j - i > 0$ we have $a^k = a^{j+(-i)} = a^j \circ (a^j)^{-1} = e$. Now for $\ell \in \mathbb{Z}$ arbitrary, we may write $\ell = qk + r$ for some integer q and some r with $0 \leq r < k$, by Proposition 3.1.8. Hence

$$a^\ell = a^{qk} \circ a^r = (a^k)^q \circ a^r = e^q \circ a^r = e \circ a^r = a^r.$$

This implies $\langle a \rangle = \{a^0, a^1, \ldots, a^{k-1}\}$, hence $|\langle a \rangle| \leq k$. (Warning: In the list $a^0, \ldots, a^{k-1}$ there may be repetitions, so k need not be the order of a.)   □

Proposition 4.2.7. *Let $(G, \circ, e)$ be a group, and let $a \in G$ with $\mathrm{ord}_G(a) = m$, for some $m \geq 1$. Then the following holds:*

(a) $\langle a \rangle = \{e, a, a^2, \ldots, a^{m-1}\}$.
(b) $a^i = a^j$ *if and only if $m \mid j - i$. (This implies that $a^i = a^{i \bmod m}$ for all $i \in \mathbb{Z}$.)*
(c) *The group $\langle a \rangle$ is isomorphic to $\mathbb{Z}_m = \{0, \ldots, m-1\}$ with addition modulo m via the mapping $i \mapsto a^i$, $i \in \mathbb{Z}_m$. In particular, $(a^i)^{-1} = a^{m-i}$.*

Proof. (a) In Lemma 4.2.6(b) we have seen that if $i < j$ and $a^i = a^j$ then $\mathrm{ord}_G(a) \leq j - i$. This implies that the elements $a^0, a^1, \ldots, a^{m-1}$ must be different, and hence must exhaust $\langle a \rangle$.
(b) By (a), we have $a^m \in \{a^0, \ldots, a^{m-1}\}$. If a^m were equal to a^i for some i, $1 \leq i < m$, then by Lemma 4.2.6(b) we would have $\mathrm{ord}_G(a) \leq m - i < m$, which is impossible. Hence $a^m = a^0 = e$. Now if $j - i = mq$, then $a^i = a^i \circ e^q = a^i \circ (a^m)^q = a^i \circ a^{mq} = a^{i+mq} = a^j$. Conversely, assume that $a^i = a^j$. Then $a^{j-i} = e = a^0$. Find q and r, $0 \leq r < m$, with $j - i = mq + r$. Then $e = a^{j-i} = a^{mq+r} = (a^m)^q \circ a^r = e^q \circ a^r = a^r$. Since $a^0 = e$ and $a^0, \ldots, a^{m-1}$ are distinct, this implies that $r = 0$, which means that $j - i = mq$.
(c) By (a), the mapping $h \colon \{0, \ldots, m - 1\} \ni i \mapsto a^i \in \langle a \rangle$ is a bijection. Clearly, $a^0 = e$. Now assume $0 \leq i, j < m$. Then $a^{(i+j) \bmod m} = a^{i+j} = a^i \circ a^j$, by (b). Finally, the inverse of i in $\mathbb{Z}_m$ is $m - i$, and $a^i \circ a^{m-i} = a^m = e$, by (b). Hence $(a^m)^{-1} = a^{m-i}$.   □

For later use, we note two simple, but important consequences of this proposition.

Proposition 4.2.8. *If $(G, \circ, e)$ is a finite group and $a \in G$ then $a^{|G|} = e$.*

Proof. The group $\langle a \rangle$ is a subgroup of G. Proposition 4.1.9 implies that $\mathrm{ord}_G(a) = |\langle a \rangle|$ is a divisor of $|G|$. By Proposition 4.2.7(b) this implies that $a^{|G|} = a^0 = e$. □

Theorem 4.2.9 (Euler). *If $m \geq 2$, then all elements $a \in \mathbb{Z}_m^*$ satisfy $a^{\varphi(m)} \bmod m = 1$.*

Proof. Apply Proposition 4.2.8 to the finite group $\mathbb{Z}_m^*$, which has cardinality $\varphi(m)$. □

If $m = p$ is a prime number, we have $\mathbb{Z}_p^* = \{1, 2, \ldots, p-1\}$, a set with $p-1$ elements, and the previous theorem turns into the following.

Theorem 4.2.10 (Fermat's Little Theorem). *If p is a prime number and $1 \leq a < p$, then $a^{p-1} \bmod p = 1$. (Consequently, $a^p \bmod p = a$ for all a, $0 \leq a < p$.)* □

4.2.3 Subgroups of Cyclic Groups

Now we have understood the structure of finite cyclic groups (they look like some $\mathbb{Z}_m$), we gather more information by analyzing their subgroup structure and the order of their elements. By Proposition 4.1.9 we know that if G is a finite cyclic group and H is a subgroup of G, then $|H|$ is a divisor of $|G|$. We will see that indeed there is exactly one subgroup of size d for each divisor d of m.

Lemma 4.2.11. *Assume $G = \langle a \rangle$ is a cyclic group of size m and H is a subgroup of G. Then*

(a) *H is cyclic;*
(b) *$H = \{a^0, a^z, a^{2z}, \ldots, a^{(d-1)z}\}$ for some divisor z of m and $d = m/z$;*
(c) *$H = \{b \in G \mid b^d = e\}$, for $d = m/z$ from (b).*

Proof. We know that $G = \{a^0, a^1, \ldots, a^{m-1}\}$ for $m = |G|$. If $|H| = 1$, then $H = \{e\} = \langle e \rangle$, and all claims are true for $z = m$ and $d = 1$. Thus assume that $d > 1$, and let $1 \leq z < m$ be minimal with $a^z \in H$.
(a) Now assume $a^i \in H$ is arbitrary. Write $i = qz + r$ for some r, $0 \leq r < z$. Then $a^r = a^i \circ (a^z)^{-q} \in H$. Since z was chosen minimal, this implies that $r = 0$. In other words, $i = qz$, or $a^i = (a^z)^q$. Hence $H = \langle a^z \rangle$, and (a) is proved.
(b) We only have to show that z is a divisor of m. (Then with $d = m/z$ we get $a^{dz} = a^m = e$, from which it is clear that $H = \langle a^z \rangle = \{a^0, a^z, a^{2z}, \ldots, a^{(d-1)z}\}$, a set with d distinct elements.) Let $r = \gcd(z, m)$. Then we may write $r = jz + km$ for some integers j, k, by Proposition 3.1.11. We get $a^r = (a^z)^j \circ (a^m)^k = (a^z)^j \in H$. Since z was chosen minimal in $\{1, 2, 3, \ldots\}$ with $a^z \in H$, this entails that $r = z$, or z divides m.

(c) Since m is a divisor of jzd for $0 \leq j < d$, we have $(a^{jz})^d = e$ for all elements $a^{jz} \in H$. Conversely, if $b^d = e$, for $b = a^i$, then $a^{id} = e$, and hence m is a divisor of $id = im/z$. This implies that i/z is an integer, and hence that $a^i \in H$, by (b).   □

We consider a converse of part (c) of Lemma 4.2.11.

Lemma 4.2.12. *Assume $G = \langle a \rangle$ is a cyclic group of size m and $s \geq 0$ is arbitrary. Then*

$$H_s = \{a \in G \mid a^s = e\}$$

is a subgroup of G with $\gcd(m, s)$ elements.
(In particular, every divisor s of m gives rise to the subgroup $H_s = \{a \in G \mid a^s = e\}$ of size s.)

Proof. It is a simple consequence of the subgroup criterion Lemma 4.1.6 that H_s is indeed a subgroup of G. Which elements a^i, $0 \leq i < m$, are in this subgroup? They must satisfy $(a^i)^s = e$, which means that m is a divisor of is. This is the case if and only if i is a multiple of $m/\gcd(m, s)$. Of these, there are $m/(m/\gcd(m, s)) = \gcd(m, s)$ many in $\{0, 1, \ldots, m - 1\}$.   □

As an example, consider the group $(\mathbb{Z}_{20}, +_{20}, 0)$, with generator 1. This group has six subgroups $H_d = \{a \in \mathbb{Z}_{20} \mid d \cdot a \equiv 0 \pmod{20}\}$, for d a divisor of 20. One generator of H_d is $20/d$. This yields the subgroups shown in Table 4.3.

d	$20/d$	H_d
1	20	$\{0\}$
2	10	$\{0, 10\}$
4	5	$\{0, 5, 10, 15\}$
5	4	$\{0, 4, 8, 12, 16\}$
10	2	$\{0, 2, 4, 6, 8, 10, 12, 14, 16, 18\}$
20	1	$\{0, 1, 2, 3, \ldots, 19\}$

Table 4.3. Subgroups of $(\mathbb{Z}_{20}, +_{20})$ and their orders

Lemma 4.2.13. *Let $G = \langle a \rangle$ be a cyclic group of size m. Then we have:*

(a) *If $b \in G$, then $\mathrm{ord}_G(b)$ is a divisor of m.*
(b) *The order of $a^i \in G$ is $m/\gcd(i, m)$.*
(c) *For each divisor d of m, G contains exactly $\varphi(d)$ elements of order d.*

Proof. (a) $\mathrm{ord}_G(b) = |\langle b \rangle|$ is a divisor of $|G|$, by Proposition 4.1.9.

(b) Assume a^i has order d. Then $a^{id} = (a^i)^d = e$, but $a^i, a^{2i}, \ldots, a^{(d-1)i}$ are different from e. By Proposition 4.2.7(b) this means that d is the smallest number $k \geq 1$ such that m divides ki. Write $i' = i/\gcd(i,m)$ and $m' = m/\gcd(i,m)$. Then

$$m \mid ki \Leftrightarrow m' \mid ki' \Leftrightarrow m' \mid k,$$

since m' and i' are relatively prime. The smallest $k \geq 1$ that is divisible by m' is $m' = m/\gcd(i,m)$ itself.

(c) By (b), we need to count the numbers $i \in \{0, 1, \ldots, m-1\}$ such that $d = m/\gcd(i,m)$, or $\gcd(i,m) = m/d$. Only numbers i of the form $j \cdot (m/d)$, $0 \leq j < d$, can have this property. Now $\gcd(j(m/d), m) = (m/d) \cdot \gcd(j, d)$ equals m/d if and only if $\gcd(j, d) = 1$. Thus, exactly the numbers $j \cdot (m/d)$, $0 \leq j < d$, with $\gcd(j, d) = 1$ are as required. There are exactly $\varphi(d)$ of them. $\square$

As an example, we consider the group $\mathbb{Z}_{25}^*$. This group of size $\varphi(25) = 20$ is cyclic with 2 as generator, since the powers $2^i \bmod m$, $0 \leq i < 20$, in this order, are $1, 2, 4, 8, 16, 7, 14, 3, 6, 12, 24, 23, 21, 17, 9, 18, 11, 22, 19, 13$.

d	$\varphi(d)$	elements of order d
1	1	$\{2^{20}\} = \{2^0\} = \{1\}$
2	1	$\{2^{10}\} = \{24\}$
4	2	$\{2^5, 2^{15}\} = \{7, 18\}$
5	4	$\{2^4, 2^8, 2^{12}, 2^{16}\} = \{16, 6, 21, 11\}$
10	4	$\{2^2, 2^6, 2^{14}, 2^{18}\} = \{4, 14, 9, 19\}$
20	8	$\{2^1, 2^3, 2^7, 2^9, 2^{11}, 2^{13}, 2^{17}, 2^{19}\} = \{2, 8, 3, 12, 23, 17, 22, 13\}$

Table 4.4. The elements of $\mathbb{Z}_{25}^*$ and their orders

4.3 Rings and Fields

Definition 4.3.1. *A **monoid** is a set M together with a binary operation $\circ$ on M with the following properties:*

(i)   *(Associativity) $(a \circ b) \circ c = a \circ (b \circ c)$, for all $a, b, c \in M$.*

(ii)   *(Neutral element) There is an element $e \in M$ that satisfies $a \circ e = e \circ a = a$ for each $m \in M$. (In particular, M is not empty.)*

*A monoid $(M, \circ, 1)$ is called **commutative** if all $a, b \in M$ satisfy $a \circ b = b \circ a$.*

An elementary and important example of a monoid is the set $\mathbb{N}$ of natural numbers with the addition operation. The neutral element is the number 0. Note that also the set $\mathbb{N}$ with the multiplication operation is a monoid, with neutral element 1.

Definition 4.3.2. *A **ring** (**with** 1) is a set R together with two binary operations $\oplus$ and $\odot$ on R and two distinct elements 0 and 1 of R with the following properties:*

(i) $(R, \oplus, 0)$ *is an abelian group (the **additive group** of the ring);*
(ii) $(R, \odot, 1)$ *is a monoid (the **multiplicative monoid** of the ring);*
(iii) (**Distributive law**) *For all $a, b, c \in R$: $(a \oplus b) \odot c = (a \odot c) \oplus (b \odot c)$.*

In short, we write $(R, \oplus, \odot, 0, 1)$ for such a ring.
*If $(R, \odot, 1)$ is a commutative monoid, the ring (with 1) is called **commutative**.*

Notation. In this text, we are dealing exclusively with **commutative rings with** 1. For convenience, we call these structures simply **rings**. (The reader should be aware that in different contexts "ring" is a wider concept.)

Proposition 4.3.3. *If $m \geq 2$ is an integer, then the structure $\mathbb{Z}_m = \{0, 1, \ldots, m-1\}$ with the binary operations*

$$a \oplus b = (a + b) \bmod m \quad and \quad a \odot b = (a \cdot b) \bmod m,$$

for which the numbers 0 and 1 are neutral elements, is a ring.

Proof. We just have to check the basic rules of operation of modular addition and multiplication:

- $(a + b) \bmod m = (b + a) \bmod m$.
- $((a + b) \bmod m + c) \bmod m = (a + (b + c) \bmod m) \bmod m$.
- Existence of inverses: $(a + (m - a)) \bmod m = 0$.
- $(a + 0) \bmod m = (0 + a) \bmod m = a$.
- $(a \cdot b) \bmod m = (b \cdot a) \bmod m$.
- $(((a \cdot b) \bmod m) \cdot c) \bmod m = (a \cdot (b \cdot c) \bmod m) \bmod m$.
- $(a \cdot 1) \bmod m = (1 \cdot a) \bmod m = a$.
- $(a \cdot (b + c) \bmod m) \bmod m = ((a \cdot b \bmod m) + (a \cdot c \bmod m)) \bmod m$.

The straightforward proofs are left to the reader. □

Definition 4.3.4. (a) *If $(R, \oplus, \odot, 0, 1)$ is a ring, then we let*

$$R^* = \{a \in R \mid there\ is\ some\ b \in R\ with\ a \odot b = 1 \};$$

*the elements of R^* are called the **units** of R.*
(b) *An element $a \in R - \{0\}$ is called a **zero divisor** if there is some $c \in R - \{0\}$ such that $a \odot c = 0$ in R.*

It is an easy exercise to show that $(R^*, \odot, 1)$ is an abelian group. Note that R^* and the set of zero divisors are disjoint: if $a \odot b = 1$ and $a \odot c = 0$, then $c = c \odot (a \odot b) = (c \odot a) \odot b = 0 \odot b = 0$.

Definition 4.3.5. *A **field** is a set F together with two binary operations $\oplus$ and $\odot$ on F and two distinct elements 0 and 1 of F with the following properties:*

(i) $(F, \oplus, \odot, 0, 1)$ *is a ring;*
(ii) $(F - \{0\}, \odot, 1)$ *is an abelian group* (*the **multiplicative group** of the field*), *denoted by F^*.*

In short, we write $(F, \oplus, \odot, 0, 1)$ for such a field.

In fields, all rules for addition, multiplication, subtraction, and division apply that we know to hold in the fields $\mathbb{R}$ and $\mathbb{Q}$. Here, we do not prove these rules systematically, but simply use them. Readers who worry about the admissibility of one or other transformation are referred to algebra texts that develop the rules for computation in fields more systematically.

The inverse of $a \in F$ in the additive group is denoted by $\ominus a$, the inverse of $a \in F^*$ in the multiplicative group is denoted by a^{-1}. The binary operation $\ominus$ is defined by $a \ominus b = a \oplus (\ominus b)$; the binary operation $\oslash$ by $a \oslash b = a \odot b^{-1}$, for $a \in F$, $b \in F^*$.

Example 4.3.6. Some infinite fields are well known, viz., the rational numbers $\mathbb{Q}$, the real numbers $\mathbb{R}$, and the complex numbers $\mathbb{C}$, with the standard operations.

In this book, however, *finite* fields are at the center of interest. The simplest finite fields are obtained by considering $\mathbb{Z}_p$ for a prime number p.

Proposition 4.3.7. *Let $m \geq 2$ be an integer. Then the following are equivalent:*

(i)   *The ring $\mathbb{Z}_m = \{0, 1, \ldots, m - 1\}$ is a field.*
(ii)   *m is a prime number.*

Proof. "(i) $\Rightarrow$ (ii)":   If m is not a prime number, we can write $r \cdot s = m \equiv 0$ (mod m) with $2 \leq r, s < m$. This means that $\{1, \ldots, m - 1\}$ is not closed under multiplication modulo m; in particular, this set does not form a group under this operation.
"(ii) $\Rightarrow$ (i)":   Conversely, assume that m is a prime number. Then $\mathbb{Z}_m^* = \{1, \ldots, m-1\}$, since no number of the latter set can have a nontrivial common factor with m. We have seen in Proposition 3.3.8 that $\mathbb{Z}_m^*$ is a group with respect to multiplication modulo m for every integer $m \geq 2$, so this is also true for the prime number m. $\square$

Note that in the case where m is a prime number, and $0 < a < m$, an inverse of a, i.e., a number x that satisfies $x \cdot a \equiv 1$ (mod m), can be calculated using the Extended Euclidean Algorithm 3.2.4 (see the remarks after Proposition 3.3.8).

We illustrate these observations by little numerical examples. $\mathbb{Z}_{12} = \{0, 1, \ldots, 11\}$ with arithmetic modulo 12 is not a field, since, for example, $3 \cdot 4 = 12 \equiv 0 \pmod{12}$, and hence $\{1, \ldots, 11\}$ is not closed under multiplication. On the other hand, $\mathbb{Z}_{13}$ is a field. We find the multiplicative inverse of 6 by applying the Extended Euclidean Algorithm to 6 and 13, which shows that $(-2) \cdot 6 + 1 \cdot 13 = 1$, from which we get that $(-2) \bmod 13 = 11$ is an inverse of 6 modulo 13.

To close the section, we note that monoids really are the natural structures in which to carry out fast exponentiation.

Proposition 4.3.8. *Let $(M, \circ, 1)$ be a monoid. There is an algorithm that for every $a \in M$ and $n \geq 0$ computes a^n in M with $O(\log n)$ multiplications in M.*

Proof. We use Algorithm 2.3.3 in the formulation for monoids M:

Algorithm 4.3.9 (Fast Modular Exponentiation in Monoids)

INPUT: Element a of monoid $(M, \circ, 1)$ and $n \geq 0$.
METHOD:

```
0      s, c: M;  u: integer;
1          u ← n;
2          s ← a;
3          c ← 1;
4      while u ≥ 1 repeat
5              if u is odd then c ← c ∘ s;
6              s ← s · s mod m;
7              u ← u div 2;
8      return c;
```

The analysis is exactly the same as that for Algorithm 2.3.3. On input a and n it carries out no more than $2\|n\| = O(\log n)$ multiplications of elements of M, and the result is correct. $\qquad\Box$

Of course, if the elements of M are structured elements (like polynomials), then the total cost of carrying out Algorithm 4.3.9 is $O(\log n)$ multiplied with the cost of one such multiplication.

4.4 Generators in Finite Fields

In this section we shall establish the basic fact that the multiplicative groups in finite fields are cyclic.

Example 4.4.1. In the field $\mathbb{Z}_{19}$, consider the powers $g^0, g^1, \ldots, g^{17}$ of the element $g = 2$ (of course, all calculations are modulo 19):

$$1, 2, 4, 8, 16, 13, 7, 14, 9, 18, 17, 15, 11, 3, 6, 12, 5, 10.$$

This sequence exhausts the whole multiplicative group $\mathbb{Z}_{19}^* = \{1, 2, \ldots, 18\}$. This means that $\mathbb{Z}_{19}^*$ is a cyclic group with generator 2.

It is the purpose of this section to show that in *every finite field F* the multiplicative group is cyclic. A generating element g of this group is called a **generator** for F. If F happens to be a field $\mathbb{Z}_p$ for a prime number p, a generator for $\mathbb{Z}_p$ is also called a **primitive element modulo** p. (Thus, 2 is a primitive element modulo 19.)

As a preparation, we need a lemma concerning Euler's totient function φ. The following example (from [21]) should make this lemma appear "obvious". Consider the 12 fractions with denominator 12 and numerator in $\{1, \ldots, 12\}$:

$$\frac{1}{12}, \frac{2}{12}, \frac{3}{12}, \frac{4}{12}, \frac{5}{12}, \frac{6}{12}, \frac{7}{12}, \frac{8}{12}, \frac{9}{12}, \frac{10}{12}, \frac{11}{12}, \frac{12}{12}.$$

Now reduce these fractions to lowest terms, by dividing numerator and denominator by their greatest common divisor:

$$\frac{1}{12}, \frac{1}{6}, \frac{1}{4}, \frac{1}{3}, \frac{5}{12}, \frac{1}{2}, \frac{7}{12}, \frac{2}{3}, \frac{3}{4}, \frac{5}{6}, \frac{11}{12}, \frac{1}{1},$$

and group them according to their denominators:

$$\frac{1}{1}; \quad \frac{1}{2}; \quad \frac{1}{3}, \frac{2}{3}; \quad \frac{1}{4}, \frac{3}{4}; \quad \frac{1}{6}, \frac{5}{6}; \quad \frac{1}{12}, \frac{5}{12}, \frac{7}{12}, \frac{11}{12}.$$

It is immediately clear that the denominators are just the divisors $1, 2, 3, 4, 6,$ 12 of 12, and that there are exactly $\varphi(d)$ fractions with denominator d, for d a divisor of 12, viz., those fractions $\frac{i}{d}$, $1 \le i \le d$, with i and d relatively prime. Since we started with 12 fractions, we have

$$\varphi(1) + \varphi(2) + \varphi(3) + \varphi(4) + \varphi(6) + \varphi(12) = 12.$$

More generally, we can show the corresponding statement for every number n in place of 12.

Lemma 4.4.2. *For every $n \in \mathbb{N}$ we have*

$$\sum_{d \mid n} \varphi(d) = n.$$

Proof. Consider the sequence

$$(a_i, b_i) = \left(\frac{i}{\gcd(i, n)}, \frac{n}{\gcd(i, n)} \right), \quad 1 \le i \le n.$$

Then each b_i is a divisor of n. Further, for each divisor d of n the pair (j, d) appears in the sequence if and only if $1 \le j \le d$ and j and d are relatively prime. Hence there are exactly $\varphi(d)$ indices i with $b_i = d$. Summing up, we obtain $\sum_{d \mid n} \varphi(d) = n$, as claimed. $\qquad \square$

Theorem 4.4.3. *If F is a finite field, then F^* is a cyclic group. In other words, there is some $g \in F^*$ with $F^* = \{1, g, g^2, \ldots, g^{|F|-2}\}$.*

Proof. Let $q = |F|$. Then $|F^*| = |F - \{0\}| = q - 1$.

For each divisor d of $q - 1$, let

$$B_d = \{b \in F^* \mid \operatorname{ord}_{F^*}(b) = d\}.$$

Claim: $|B_d| = 0$ or $|B_d| = \varphi(d)$.

Proof of Claim: Assume $B_d \neq \emptyset$, and choose some element a of B_d. By Proposition 4.2.7, this element generates the subgroup $\langle a \rangle = \{a^0, a^1, \ldots, a^{d-1}\}$ of size d. Clearly, for $0 \leq i < d$ we have $(a^i)^d = (a^d)^i = 1$. Hence each of the d elements of $\langle a \rangle$ is a root of the polynomial $X^d - 1$ in F. We now allow ourselves to use Theorem 7.5.1, to be proved later in Sect. 7.5, to note that the polynomial $X^d - 1$ does not have more than d roots in F, hence $\langle a \rangle$ comprises exactly the set of roots of $X^d - 1$. Since each element b of B_d satisfies $b^d = 1$, hence is a root of $X^d - 1$, we obtain $B_d \subseteq \langle a \rangle$. Now applying Proposition 4.2.13(c) we note that $\langle a \rangle$ contains exactly $\varphi(d)$ elements of order d in $\langle a \rangle$, which is the same as the order in F^*. Thus, $|B_d| = \varphi(d)$, and the claim is proved.

By Proposition 4.1.9, the order of each element $a \in F^*$ is a divisor of $|F^*| = q - 1$. Thus, the sets B_d, $d \mid q - 1$, form a partition of F^* into disjoint subsets. Hence we have

$$q - 1 = |F^*| = \sum_{d \mid q-1} |B_d|. \tag{4.4.1}$$

If we apply Lemma 4.4.2 to $q - 1$, we obtain

$$q - 1 = \sum_{d \mid q-1} \varphi(d). \tag{4.4.2}$$

Combining (4.4.1) and (4.4.2) with the claim, we see that in fact *none* of the B_d's can be empty. In particular $B_{q-1} \neq \emptyset$, and each of the $\varphi(q-1)$ elements $g \in B_{q-1}$ is a primitive element of F. $\qquad\square$

Corollary 4.4.4. *If p is a prime number, then $\mathbb{Z}_p^*$ is a cyclic group with $\varphi(p-1)$ generators (called "**primitive elements modulo p**").* $\qquad\square$

Example 4.4.5. The $\varphi(12) = 4$ primitive elements modulo 13 are 2, 6, 7, and 11.

Definition 4.4.6. *If p is a prime number, and n is an integer not divisible by p, we write $\operatorname{ord}_p(n)$ for $\operatorname{ord}_{\mathbb{Z}_p^*}(n \bmod p)$, and call this number the **order of n modulo p**.*

Clearly, $\operatorname{ord}_p(n)$ is the smallest $i \geq 1$ that satisfies $n^i \bmod p = 1$, and $\operatorname{ord}_p(n)$ is a divisor of $|\mathbb{Z}_p^*| = p - 1$.

5. The Miller-Rabin Test

In this chapter, we describe and analyze our first randomized primality test, which admits extremely efficient implementations and a reasonable worst-case error probability of $\frac{1}{4}$. The error analysis employs simple group theory and the Chinese Remainder Theorem.

5.1 The Fermat Test

Recall Fermat's Little Theorem (Theorem 4.2.10), which says that if p is a prime number and $1 \leq a < p$, then $a^{p-1} \bmod p = 1$.

We may use Fermat's Little Theorem as a means for identifying composite numbers. Let us take $a = 2$, and for given n, calculate $2^{n-1} \bmod n$. (By using fast exponentiation, this has cost $O((\log n)^3)$.) We start: $2^2 \bmod 3 = 1$, $2^3 \bmod 4 = 8 \bmod 4 = 0$, $2^4 \bmod 5 = 16 \bmod 5 = 1$, $2^5 \bmod 6 = 32 \bmod 6 = 2$. The values for some small n are given in Table 5.1. "Obviously" this is

n	3	4	5	6	7	8	9	10	11	12	13	14	15	16	17
$2^{n-1} \bmod n$	1	0	1	2	1	0	4	2	1	8	1	2	4	0	1

Table 5.1. Primality test by calculating $2^{n-1} \bmod n$

a very good primality criterion — for prime numbers $n \leq 17$ we get 1, and for nonprimes we get some value different from 1. By Theorem 4.2.10, if $2^{n-1} \bmod n \neq 1$ we have a definite certificate for the fact that n is composite. We then call 2 a *Fermat witness* for n (more exactly, a witness for the fact that n is composite). Of course, nothing is special about the base 2 here, and we define more generally:

Definition 5.1.1. *A number a, $1 \leq a < n$, is called an **F-witness** for n if $a^{n-1} \bmod n \neq 1$.*

If n has an F-witness, it is composite. It is important to note that an F-witness a for n is a certificate for the compositeness of n, but it does not reveal any information about possible factorizations of n. We will see more such

M. Dietzfelbinger: Primality Testing in Polynomial Time, LNCS 3000, pp. 73-84, 2004.
© Springer-Verlag Berlin Heidelberg 2004

certificate systems later on. With some patience (or a computer program) one can check that 2 is an F-witness for all composite numbers not exceeding 340, but for the composite number $341 = 11 \cdot 31$ we get $2^{340} \bmod 341 = 1$. As far as one can tell from looking at the value $2^{340} \bmod 341$, there is no indication that 341 is not prime. We call 2 a *Fermat liar* for 341 in the sense of the following definition.

Definition 5.1.2. *For an odd composite number n we call an element a, $1 \le a \le n - 1$, an **F-liar** if $a^{n-1} \bmod n = 1$.*

Note that 1 and $n - 1$ trivially are F-liars for all odd composite n, since 1^{n-1} $\bmod n = 1$ in any case, and $(n - 1)^{n-1} \equiv (-1)^{n-1} \equiv 1 \pmod{n}$, since $n - 1$ is even. Continuing the example from above, we have $3^{340} \bmod 341 = 56$, so 3 is an F-witness for 341. (For much more information on composite numbers for which 2 is an F-liar [so-called pseudoprimes base 2], see [33].)

Let us note that some kind of reverse of Fermat's Little Theorem is true.

Lemma 5.1.3. *Let $n \ge 2$ be an integer.*

(a) *If $1 \le a < n$ satisfies $a^r \bmod n = 1$ for some $r \ge 1$, then $a \in \mathbb{Z}_n^*$.*
(b) *If $a^{n-1} \bmod n = 1$ for all a, $1 \le a < n$, then n is a prime number.*

Proof. (a) If $a^r \bmod n = 1$, then $a \cdot a^{r-1} \bmod n = 1$, hence $a \in \mathbb{Z}_n^*$, by Proposition 3.3.8(c).
(b) Assume $a^{n-1} \bmod n = 1$ for all a, $1 \le a < n$. From (a) it follows that then $\mathbb{Z}_n^* = \{1, \ldots, n - 1\}$, and this is the same as to say that n is a prime number.   □

By Lemma 5.1.3(b) there will always be *some* F-witnesses for an odd composite number n. More precisely, the $n - 1 - \varphi(n)$ elements of

$$\{1, \ldots, n - 1\} - \mathbb{Z}_n^* = \{a \mid 1 \le a < n, \gcd(a, n) > 1\}$$

cannot satisfy $a^{n-1} \bmod n = 1$. Unfortunately, for many composite numbers n this set is very slim. Just assume that n is a product of two distinct primes p and q. Then a satisfies $\gcd(a, n) > 1$ if and only if $p \mid a$ or $q \mid a$. There are exactly $p + q - 2$ such numbers in $\{1, \ldots, n - 1\}$, which is very small in comparison to n if p and q are roughly equal. Let us look at an example: $n = 91 = 7 \cdot 13$. Table 5.2 shows that there are 18 multiples of 7 and 13 (for larger p and q the fraction of these "forced" F-witnesses will be smaller), and, apart from these, 36 F-witnesses and 36 F-liars in $\{1, 2, \ldots, 90\}$. In this example there are some more F-witnesses than F-liars. If this were the case for all odd composite numbers n, it would be a great strategy to just grope at random for some a that satisfies $a^{n-1} \bmod n \ne 1$.

This leads us to our first attempt at a randomized primality test.

multiples of 7	7, 14, 21, 28, 35, 42, 49, 56, 63, 70, 77, 84
multiples of 13	13, 26, 39, 52, 65, 78
F-witnesses in $\mathbb{Z}_{91}^*$	2, 5, 6, 8, 11, 15, 18, 19, 20, 24, 31, 32, 33, 34, 37, 41, 44, 45, 46, 47, 50, 54, 57, 58, 59, 60, 67, 71, 72, 73, 76, 80, 83, 85, 86, 89
F-liars	1, 3, 4, 9, 10, 12, 16, 17, 22, 23, 25, 27, 29, 30, 36, 38, 40, 43, 48, 51, 53, 55, 61, 62, 64, 66, 68, 69, 74, 75, 79, 81, 82, 87, 88, 90

Table 5.2. F-witnesses and F-liars for $n = 91 = 7 \cdot 13$

Algorithm 5.1.4 (Fermat Test)

INPUT: Odd integer $n \geq 3$.
METHOD:
1   Let a be randomly chosen from $\{2, \ldots, n - 2\}$;
2   **if** $a^{n-1} \bmod n \neq 1$
3   **then return** 1;
4   **else return** 0;

The analysis of the running time is obvious: The most expensive part is the calculation of $a^{n-1} \bmod n$ by fast exponentiation, which according to Lemma 2.3.4 takes $O(\log n)$ arithmetic operations and $O((\log n)^3)$ bit operations. Further, it is clear that if the algorithm outputs 1, then it has detected an F-witness a for n, hence n is guaranteed to be composite. For $n = 91$, the misleading result 0 is obtained if the random choice for a hits one of the 34 F-liars other than 1 and 90, which has probability $\frac{34}{88} = \frac{17}{44}$.

With a little group theory it is easy to see that for many composite numbers n there will be an abundance of F-witnesses, so that this simple test will succeed with constant probability.

Theorem 5.1.5. *If $n \geq 3$ is an odd composite number such that there is at least one F-witness a in $\mathbb{Z}_n^*$, then the Fermat test applied to n gives answer 1 with probability more than $\frac{1}{2}$.*

Proof. In Lemma 5.1.3(a) we have seen that the set

$$L_n^{\mathrm{F}} = \{a \mid 1 \leq a < n,\ a^{n-1} \bmod n = 1\}$$

of F-liars for n is a subset of $\mathbb{Z}_n^*$. We now show that L_n^{F} even is a subgroup of $\mathbb{Z}_n^*$. Since $\mathbb{Z}_n^*$ is a finite group, it is sufficient to check the following conditions (see Lemma 4.1.6):

(i) $1 \in L_n^F$, since $1^{n-1} = 1$;

(ii) L_n^F is closed under the group operation in $\mathbb{Z}_n^*$, which is multiplication modulo n, since if $a^{n-1} \bmod n = 1$ and $b^{n-1} \bmod n = 1$, then $(ab)^{n-1} \equiv a^{n-1} \cdot b^{n-1} \equiv 1 \cdot 1 \equiv 1 \pmod{n}$.

By the assumption that there is at least one F-witness in $\mathbb{Z}_n^*$ we get that L_n^F is a *proper* subgroup of $\mathbb{Z}_n^*$. This observation gives us a much stronger bound on $|L_n^F|$ than just $\varphi(n) - 1$. Namely, by Proposition 4.1.9 the size of L_n^F must be a proper divisor of $\varphi(n) < n - 1$, in particular $|L_n^F| \leq (n-2)/2$. Thus, the probability that an a randomly chosen from $\{2, \ldots, n-2\}$ is in $L_n^F - \{1, n-1\}$ is at most

$$\frac{(n-2)/2 - 2}{n-3} = \frac{n-6}{2(n-3)} < \frac{1}{2}, \tag{5.1.1}$$

as desired.   □

Of course, an algorithm that gives a wrong answer with probability up to $\frac{1}{2}$ should not be trusted. More convincing success probabilities may be obtained by repeating the Fermat test, as described next.

Algorithm 5.1.6 (Iterated Fermat Test)

INPUT: Odd integer $n \geq 3$, integer $\ell \geq 1$.
METHOD:
```
1      repeat ℓ times
2          a ← a randomly chosen element of {2, ..., n − 2};
3          if aⁿ⁻¹ mod n ≠ 1 then return 1;
4      return 0;
```

Again we note that if the output is 1, then the algorithm has found an F-witness for n, hence n is composite. Turned the other way round, if n is prime the output is guaranteed to be 0. On the other hand, if n is composite, and we are in the situation of the previous theorem — there is at least one F-witness a with $\gcd(a, n) = 1$ — then the probability that in all ℓ attempts an F-liar a is chosen is smaller than $\left(\frac{1}{2}\right)^\ell = 2^{-\ell}$. Thus, by choosing ℓ large enough, this error probability can be made as small as desired.

We remark that if random numbers n are to be tested for primality, then the Fermat test is a very efficient and reliable method. (For details see [17] and [14].) Unfortunately, there are some, presumably rare, stubborn composite numbers that do not yield to the Fermat test, because all elements of $\mathbb{Z}_n^*$ are F-liars.

Definition 5.1.7. *An odd composite number n is called a **Carmichael number** if $a^{n-1} \bmod n = 1$ for all $a \in \mathbb{Z}_n^*$.*

The smallest Carmichael number is $561 = 3 \cdot 11 \cdot 17$. Only in 1994 was it shown that there are infinitely many Carmichael numbers, and an asymptotic lower bound for their density was given: there is some x_0 such that for all $x \geq x_0$

the set $\{n \mid n \leq x\}$ contains more than $x^{2/7}$ Carmichael numbers [5]. If a Carmichael number is fed into the Fermat test, the probability that the wrong answer 0 is given is

$$\frac{\varphi(n) - 2}{n - 3} > \frac{\varphi(n)}{n} = \prod_{\substack{p \text{ prime} \\ p \mid n}} \left(1 - \frac{1}{p}\right).$$

(The last equality is from Proposition 3.5.12.) This bound is annoyingly close to 1 if n has only few and large prime factors. For example, in lists of Carmichael numbers generated by computer calculations ([32] and associated website) one finds Carmichael numbers like $n = 651693055693681 = 72931 \cdot 87517 \cdot 102103$, with $\varphi(n)/n > 0.99996$. The repetition trick does not help here either, since if the smallest prime factor of a Carmichael number n is p_0, and n has only 3 or 4 factors, then $\Omega(p_0)$ repetitions are necessary to make the error probability smaller than $\frac{1}{2}$. This is unfeasible as soon as p_0 has more than 20 decimal digits, say.

Thus for a reliable primality test that works for *all* composite numbers, we have to go beyond the Fermat test. Before we formulate this more clever test, we state and prove a basic property of Carmichael numbers.

Lemma 5.1.8. *If n is a Carmichael number, then n is a product of at least three distinct prime factors.*

Proof. We prove the contraposition: Assume that n is not a product of three or more distinct primes. We exhibit an F-witness a in $\mathbb{Z}_n^*$. For this, we consider two cases.

Case 1: n is divisible by p^2 for some prime number $p \geq 3$. — Write $n = p^k \cdot m$ for some $k \geq 2$ and some m that is not divisible by p. If $m = 1$, let $a = 1 + p$. If $m \geq 3$, then by the Chinese Remainder Theorem 3.4.1 we may choose some a, $1 \leq a < p^2 \cdot m \leq n$, with

$$a \equiv 1 + p \quad (\text{mod } p^2) \quad \text{and}$$
$$a \equiv 1 \quad (\text{mod } m).$$

We claim that a is an F-witness in $\mathbb{Z}_n^*$. Why is a in $\mathbb{Z}_n^*$? Since p^2 divides $a - (1 + p)$ in both cases, p does not divide a. Further, $\gcd(a, m) = 1$. (If $m = 1$, this is trivial; if $m \geq 3$, it follows from $a \equiv 1 \pmod{m}$.) Thus, $\gcd(a, n) = 1$, as desired.

Next we show that a is an F-witness for n. Assume for a contradiction that $a^{n-1} \equiv 1 \pmod{n}$. Since $p^2 \mid n$, we get $a^{n-1} \equiv 1 \pmod{p^2}$. On the other hand, by the binomial theorem,

$$a^{n-1} \equiv (1+p)^{n-1} \equiv 1 + (n-1)p + \sum_{2 \leq i \leq n-1} \binom{n-1}{i} p^i \equiv 1 + (n-1)p \pmod{p^2}.$$

Thus $(n - 1)p \equiv 0 \pmod{p^2}$, which means that p^2 divides $(n - 1)p$. But this is impossible, since p does not divide $n - 1 = p^k \cdot m - 1$.

Case 2: $n = p \cdot q$ for two distinct prime numbers p and q. — We may arrange the factors so that $p > q$. Again, we construct an F-witness a in $\mathbb{Z}_n^*$, as follows. We know (by Theorem 4.4.3) that the group $\mathbb{Z}_p^*$ is cyclic, i.e., it has a generator g. By the Chinese Remainder Theorem 3.4.1, we may choose an element a, $1 \leq a < n$, such that

$$a \equiv g \pmod{p} \quad \text{and}$$
$$a \equiv 1 \pmod{q}.$$

The element a is divisible by neither p nor q, hence $a \in \mathbb{Z}_n^*$. Now assume for a contradiction that $a^{n-1} \bmod n = 1$. Since p divides n, this entails

$$g^{n-1} \bmod p = a^{n-1} \bmod p = 1.$$

Since g has order $p - 1$ in $\mathbb{Z}_p^*$, we conclude, by Proposition 4.2.7(b), that $p - 1$ divides $n - 1$. Now $n - 1 = pq - 1 = (p - 1)q + q - 1$, so we obtain that $p - 1$ divides $q - 1$, in particular, $p \leq q$, which is the desired contradiction. □

5.2 Nontrivial Square Roots of 1

We consider another property of arithmetic modulo p for a prime number p that can be used as a certificate for compositeness.

Definition 5.2.1. *Let* $1 \leq a < n$. *Then* a *is called a* **square root of** 1 **modulo** n *if* $a^2 \bmod n = 1$.

In the situation of this definition, the numbers 1 and $n - 1$ are always square roots of 1 modulo n (indeed, $(n - 1)^2 \equiv (-1)^2 \equiv 1 \pmod{n}$); they are called the *trivial* square roots of 1 modulo n. If n is a prime number, there are no other square roots of 1 modulo n.

Lemma 5.2.2. *If* p *is a prime number and* $1 \leq a < p$ *and* $a^2 \bmod p = 1$, *then* $a = 1$ *or* $a = p - 1$.

Proof. We have $(a^2 - 1) \bmod p = (a + 1)(a - 1) \bmod p = 0$; that means, p divides $(a+1)(a-1)$. Since p is prime, p divides $a+1$ or $a-1$, hence $a = p-1$ or $a = 1$. □

Thus, if we find some nontrivial square root of 1 modulo n, then n is certainly composite.

For example, all the square roots of 1 modulo 91 are 1, 27, 64, and 90. More generally, using the Chinese Remainder Theorem 3.4.3 it is not hard to see that if $n = p_1 \cdots p_r$ for distinct odd primes $p_1, \ldots, p_r$, then there are exactly 2^r square roots of 1 modulo n, namely those numbers a, $0 \leq a < n$, that satisfy $a \bmod p_j \in \{1, p_j - 1\}$, for $1 \leq j \leq r$. This means that unless

n has extremely many prime factors, it is useless to try to find nontrivial square roots of 1 modulo n by testing randomly chosen a.

Instead, we go back to the Fermat test. Let us look at a^{n-1} mod n a little more closely. Of course, we are only interested in odd numbers n. Then $n-1$ is even, and can be written as $n - 1 = u \cdot 2^k$ for some odd number u and some $k \geq 1$. Thus, $a^{n-1} \equiv ((a^u) \bmod n)^{2^k} \bmod n$, which means that we may calculate a^{n-1} mod n with $k + 1$ intermediate steps: if we let

$$b_0 = a^u \bmod n; \ b_i = (b_{i-1}^2) \bmod n, \text{ for } i = 1, \ldots, k,$$

then $b_k = a^{n-1}$ mod n. For example, for $n = 325 = 5^2 \cdot 13$ we get $n - 1 = 324 = 81 \cdot 2^2$. In Table 5.3 we calculate the powers a^{81}, a^{162}, and a^{324}, all modulo 325, for several a.

a	$b_0 = a^{81}$	$b_1 = a^{162}$	$b_2 = a^{324}$
2	252	129	66
7	307	324	1
32	57	324	1
49	324	1	1
65	0	0	0
126	1	1	1
201	226	51	1
224	274	1	1

Table 5.3. Powers a^{n-1} mod n calculated with intermediate steps, $n = 325$

We see that 2 is an F-witness for 325 from $\mathbb{Z}_{325}^*$, and 65 is an F-witness not in $\mathbb{Z}_{325}^*$. In contrast, 7, 32, 49, 126, 201, and 224 are all F-liars for 325. Calculating 201^{324} mod 325 with two intermediate steps leads us to detect that 51 is a nontrivial square root of 1, which proves that 325 is not prime. Similarly, the calculation with base 224 reveals that 274 is a nontrivial square root of 1. On the other hand, the corresponding calculation with bases 7, 32, or 49 does not give any information, since $7^{162} \equiv 32^{162} \equiv -1 \pmod{325}$ and $49^{81} \equiv -1 \pmod{325}$. Similarly, calculating the powers of 126 does not reveal a nontrivial square root of 1, since 126^{81} mod $325 = 1$.

What can the sequence $b_0, \ldots, b_k$ look like in general? We first note that if $b_i = 1$ or $b_i = n - 1$, then the remaining elements $b_{i+1}, \ldots, b_k$ must all equal 1, since $1^2 = 1$ and $(n-1)^2$ mod $n = 1$. Thus in general the sequence starts with zero or more elements $\notin \{1, n-1\}$, and ends with a sequence of zero or more 1's. The two parts may or may not be separated by an entry $n - 1$. All possible patterns are depicted in Table 5.4, where "$*$" represents an arbitrary element $\notin \{1, n - 1\}$. We distinguish four cases:

b_0	b_1	$\cdots$				$\cdots$	b_{k-1}	b_k	Case
1	1	$\cdots$	1	1	1	$\cdots$	1	1	1a
$n-1$	1	$\cdots$	1	1	1	$\cdots$	1	1	1b
*	*	$\cdots$	*	$n-1$	1	$\cdots$	1	1	1b
*	*	$\cdots$	*	*	*	$\cdots$	*	$n-1$	2
*	*	$\cdots$	*	*	*	$\cdots$	*	*	2
*	*	$\cdots$	*	1	1	$\cdots$	1	1	3
*	*	$\cdots$	*	*	*	$\cdots$	*	1	3

Table 5.4. Powers a^{n-1} mod n calculated with intermediate steps, possible cases.

Case 1a: $b_0 = 1$.
Case 1b: $b_0 \neq 1$, and there is some $i \leq k - 1$ such that $b_i = n - 1$.
— In Cases 1a and 1b we certainly have that $b_k = 1$; no information about n being prime or not is gained.
Case 2: $b_k \neq 1$. — Then n is composite, since a is an F-witness for n.
Case 3: $b_0 \neq 1$, but $b_k = 1$, and $n - 1$ does not occur in the sequence $b_0, \ldots, b_{k-1}$. — Consider the minimal $i \geq 1$ with $b_i = 1$. By the assumption, $b_{i-1} \notin \{1, n-1\}$, hence b_{i-1} is a nontrivial square root of 1 modulo n. Thus n is composite in this case.

In Cases 2 and 3 the element a constitutes a certificate for the fact that n is composite. The disjunction of Cases 2 and 3 is that $b_0 \neq 1$ and that $n - 1$ does not occur in the sequence $b_0, \ldots, b_{k-1}$. Note that, surprisingly, the value of b_k becomes irrelevant when these two cases are combined. We capture this condition in the following definition.

Definition 5.2.3. *Let $n \geq 3$ be odd, and write $n = u \cdot 2^k$, u odd, $k \geq 1$. A number a, $1 \leq a < n$, is called an **A-witness for** n if a^u mod $n \neq 1$ and $a^{u \cdot 2^i}$ mod $n \neq n - 1$ for all i, $0 \leq i < k$. If n is composite and a is not an A-witness for n, then a is called an **A-liar for** n.*

Lemma 5.2.4. *If a is an A-witness for n, then n is composite.*

Proof. If a is an A-witness for n, then to the sequence $b_i = a^{u \cdot 2^i}$ mod n, $0 \leq i \leq k$, Case 2 or Case 3 of the preceding discussion applies, hence n is composite.   □

We combine this observation with the idea of choosing some a from $\{2, \ldots, n - 2\}$ at random into a strengthening of the Fermat test, called the *Miller-Rabin test*.

Historically, Artjuhov [7] had proposed considering the sequence $b_i = a^{u \cdot 2^i}$ mod n, for $0 \leq i \leq k$, for testing n for compositeness. Later, Miller [29] used the criterion in his deterministic algorithm that will have polynomial running time if the Extended Riemann Hypothesis (ERH, a number-theoretical conjecture) is true. He showed that, assuming the ERH, the smallest A-witness for a composite number n will be of size $O((\ln n)^2)$. Later, Bach [8] gave an explicit bound of $2(\ln n)^2$ for the smallest A-witness. The resulting deterministic primality test is obvious: for $a = 2, 3, \ldots, \lfloor 2(\ln n)^2 \rfloor$ check whether a is an A-witness for n. If all these a's fail to be A-witnesses, n is a prime number. The algorithm uses $O((\log n)^3)$ arithmetic operations, but its correctness hinges on the correctness of the ERH.

Afterwards, around 1980, Rabin (and independently Monier) recognized the possibility of turning Miller's deterministic search for an A-witness into an efficient *randomized* algorithm. Independently, an alternative randomized algorithm with very similar properties, but based on different number-theoretical principles, was discovered by Solovay and Strassen, see Chap. 6. Indeed, these very efficient randomized algorithms for a problem for which no efficient deterministic algorithm had been available before were the very convincing early examples of the importance and practical usefulness of randomized algorithms for discrete computational problems. We next describe the randomized version of the compositeness test based on the concept of an A-witness, now commonly called the Miller-Rabin test.

Algorithm 5.2.5 (Miller-Rabin Test)

INPUT: Odd integer $n \geq 3$.
METHOD:

```
1     Find u odd and k so that n = u · 2^k;
2     Let a be randomly chosen from {2, ..., n − 2};
3     b ← a^u mod n;
4     if b ∈ {1, n − 1} then return 0;
5     repeat k − 1 times
6          b ← b^2 mod n;
7          if b = n − 1 then return 0;
8          if b = 1 then return 1;
9     return 1;
```

Let us first analyze the running time of this algorithm. It takes at most $\log n$ divisions by 2 to find u and k (line 1). (Exploiting the fact that n is represented in binary in a typical computer makes it possible to speed this part up by using shifts of binary words instead of divisions.) The calculation of $a^u \bmod n$ by fast exponentiation in line 3 takes $O(\log n)$ arithmetic operations and $O((\log n)^3)$ (naive implementation of multiplication and division) resp. $O^\sim((\log n)^2)$ (faster implementations) bit operations; see Lemma 2.3.4. Finally, the loop in lines 5–8 is carried out $k \leq \log n$ times; in each iteration the multiplication modulo n is the most expensive operation. Overall, the algorithm uses $O(\log n)$ arithmetic operations and $O((\log n)^3)$ resp. $O^\sim((\log n)^2)$ bit operations.

We now turn to studying the output behavior.

Lemma 5.2.6. *If the Miller-Rabin test yields output 1, then n is composite.*

Proof. Let a be the element chosen in the algorithm, and assume that the output is 1. We show that a is an A-witness for n. (By Lemma 5.2.4, this implies that n is composite.) We refer to the sequence $b_i = a^{u \cdot 2^i} \bmod n$, $0 \le i \le k$, as above. Clearly, in line 3 the variable **b** is initialized to b_0, and the content of **b** changes from $b_{i-1} = a^{u \cdot 2^{i-1}} \bmod n$ to $b_i = a^{u \cdot 2^i} \bmod n$ when line 6 is carried out for the ith time. There are two ways for the output 1 to occur.

Case (a): There is some i, $1 \le i \le k-1$, such that in the course of the ith execution of line 7 the algorithm finds that $b_i = 1$. — By the test in line 4, and the tests in lines 7 and 8 carried out during the previous executions of the loop, we know that $b_0, \ldots, b_{i-1} \notin \{1, n-1\}$. This entails that $b_0, \ldots, b_{k-1}$ does not contain $n-1$, hence a is an A-witness.

Case (b): Line 9 is executed. — By the tests performed by the algorithm in line 4 and in the $k-1$ executions of lines 8 and 9, this means that $b_0, \ldots, b_{k-1}$ are all different from 1 and $n-1$. Again, a is an A-witness.   □

It remains to analyze the output behavior of the algorithm in the case that n is a composite number.

5.3 Error Bound for the Miller-Rabin Test

Throughout this section we assume that n is a fixed odd composite number. We show that the probability that Algorithm 5.2.5 gives the erroneous output 0 is smaller than $\frac{1}{2}$.

In order to bound the number of A-liars, we would like to proceed as in the proof of Theorem 5.1.5, where we showed that the F-liars form a proper subgroup of $\mathbb{Z}_n^*$. Unfortunately, the set of A-liars need not be a subgroup. For example, with $n = 325$, we find in Table 5.3 the A-liars 7 and 32, and see that their product 224 is an A-witness. We circumvent this difficulty by identifying a proper subgroup of $\mathbb{Z}_n^*$ that contains all A-liars.

If n is not a Carmichael number, this approach is easy to realize. We simply note that in this case $L_n^A \subseteq L_n^F$, and argue as in the proof of Theorem 5.1.5 to see that the fraction of A-liars in $\{2, \ldots, n-2\}$ is smaller than $\frac{1}{2}$.

From here on, let us assume that n is a Carmichael number. Our task is to find a proper subgroup B_n^A of $\mathbb{Z}_n^*$ that contains all A-liars.

Let i_0 be the maximal $i \ge 0$ such that there is some A-liar a_0 with $a_0^{u \cdot 2^i} \bmod n = n - 1$. Since u is odd, $(n-1)^u \equiv (-1)^u \equiv -1 \pmod{n}$, hence such an i exists. Since n is a Carmichael number, $a_0^{u \cdot 2^k} \bmod n = a_0^{n-1} \bmod n = 1$, hence $0 \le i_0 < k$. We define:

$$B_n^A = \{ a \mid 0 \le a < n, \ a^{u \cdot 2^{i_0}} \bmod n \in \{1, n-1\} \}, \tag{5.3.2}$$

and show that this set has the desired properties.

Lemma 5.3.1.

(a) $L_n^A \subseteq B_n^A$.
(b) B_n^A is a subgroup of $\mathbb{Z}_n^*$.
(c) $\mathbb{Z}_n^* - B_n^A \neq \emptyset$.

Proof. (a) Let a be an arbitrary A-liar.
Case 1: $a^u \bmod n = 1$. — Then $a^{u \cdot 2^{i_0}} \bmod n = 1$ as well, and hence $a \in B_n^A$.
Case 2: $a^{u \cdot 2^i} \bmod n = n - 1$, for some i. — Then $0 \leq i \leq i_0$ by the definition of i_0. Now if $i = i_0$, we directly have that $a \in B_n^A$; if $i < i_0$, then

$$a^{u \cdot 2^{i_0}} \bmod n = (a^{u \cdot 2^i} \bmod n)^{2^{i_0 - i}} \bmod n = 1,$$

and hence $a \in B_n^A$.
(b) We check the two conditions from Lemma 4.1.6:

   (i) $1 \in B_n^A$, since $1^{u \cdot 2^{i_0}} \bmod n = 1$.
   (ii) B_n^A is closed under the group operation in $\mathbb{Z}_n^*$: Let $a, b \in B_n^A$.
        Then $a^{u \cdot 2^{i_0}} \bmod n$, $b^{u \cdot 2^{i_0}} \bmod n \in \{1, n - 1\}$. Since $1 \cdot 1 = 1$,
        $1 \cdot (n - 1) = (n - 1) \cdot 1 = n - 1$, and $(n - 1) \cdot (n - 1) \bmod n = 1$,
        we have

$$(ab)^{u \cdot 2^{i_0}} \bmod n = ((a^{u \cdot 2^{i_0}} \bmod n) \cdot (b^{u \cdot 2^{i_0}} \bmod n)) \bmod n \in \{1, n-1\}.$$

   It follows that $ab \bmod n \in B_n^A$.

(c) By Lemma 5.1.8, the Carmichael number n has at least three different prime factors, and hence can be written as $n = n_1 \cdot n_2$ for odd numbers n_1, n_2 that are relatively prime.
   Recall that a_0 is an A-liar with $a_0^{u \cdot 2^{i_0}} \equiv -1 \pmod{n}$. Let $a_1 = a_0 \bmod n_1$. By the Chinese Remainder Theorem 3.4.1 there is a unique number $a \in \{0, \ldots, n - 1\}$ with

$$a \equiv a_1 \pmod{n_1} \quad \text{and} \quad a \equiv 1 \pmod{n_2}. \tag{5.3.3}$$

We show that a is an element of $\mathbb{Z}_n^* - B_n^A$.
   Calculating modulo n_1, we have that $a \equiv a_0 \pmod{n_1}$, and hence

$$a^{u \cdot 2^{i_0}} \equiv -1 \pmod{n_1}. \tag{5.3.4}$$

Calculating modulo n_2, we see that

$$a^{u \cdot 2^{i_0}} \equiv 1^{u \cdot 2^{i_0}} \equiv 1 \pmod{n_2}. \tag{5.3.5}$$

Now (5.3.4) entails that

$$a^{u \cdot 2^{i_0}} \not\equiv 1 \pmod{n}$$

and (5.3.5) entails that

$$a^{u \cdot 2^{i_0}} \not\equiv -1 \pmod{n}.$$

This means that $a^{u \cdot 2^{i_0}} \bmod n \notin \{1, n-1\}$, and hence $a \notin B_n^A$. Further, $a^{u \cdot 2^{i_0+1}} \bmod n_1 = 1$ and $a^{u \cdot 2^{i_0+1}} \bmod n_2 = 1$, hence, by the Chinese Remainder Theorem, $a^{u \cdot 2^{i_0+1}} \bmod n = 1$. By Lemma 5.1.3(a) we conclude that $a \in \mathbb{Z}_n^*$, and the proof of Lemma 5.3.1(c) is complete.   □

Example 5.3.2. Consider the number $n = 325 = 13 \cdot 25$. (For the purpose of this illustration it is not relevant that 325 is not a Carmichael number.) Going back to Table 5.3, we note that the A-liar 32 satisfies $32^{162} \bmod 325 = 324$. The unique number a with $a \bmod 13 = 32 \bmod 13 = 6$ and $a \bmod 25 = 1$ is $a = 201$. From the table, we read off that

$$201^{162} \bmod 325 = 51 \notin \{1, 324\}, \quad \text{but } 201^{324} \bmod 325 = 1.$$

(Note that $51 \equiv 1 \pmod{25}$ and $51 \equiv -1 \pmod{13}$.) In particular, with 201 we have an element of $\mathbb{Z}_{325}^*$ not in B_{325}^A.

We have shown an error bound of $\frac{1}{2}$ for the Miller-Rabin algorithm. In fact, it can be shown by a different, more involved analysis, see, for example, [16, p. 127], that the error probability is bounded by $\frac{1}{4}$. By ℓ-fold repetition, the bound on the error probability may be lowered to $4^{-\ell}$, in exactly the same manner as described in Algorithm 5.1.6. We summarize:

Proposition 5.3.3. *Algorithm 5.2.5, when applied ℓ times to an input number n, needs $O(\ell \cdot \log n)$ arithmetic operations and $O(\ell \cdot (\log n)^3)$ bit operations (simple methods) resp. $O^{\sim}(\ell \cdot (\log n)^2)$ bit operations (best methods). If n is a prime number, the output is 0, if n is composite, the probability that output 0 is given is smaller than $4^{-\ell}$.*   □

6. The Solovay-Strassen Test

The primality test of Solovay and Strassen [39] is similar in flavor to the Miller-Rabin test. Historically, it predates the Miller-Rabin test. Like the Miller-Rabin test it is a randomized procedure; it is capable of recognizing composite numbers with a probability of at least $\frac{1}{2}$. To explain how the test works, we must define quadratic residues and introduce the Legendre symbol and the Jacobi symbol. For efficient evaluation of these quantities the Quadratic Reciprocity Law is central.

6.1 Quadratic Residues

For reasons of convention, we introduce special notation for the squares in the multiplicative group $\mathbb{Z}_m^*$.

Definition 6.1.1. *For $m \geq 2$ and $a \in \mathbb{Z}$ with $\gcd(a, m) = 1$ we say that a is a **quadratic residue** modulo m if $a \equiv x^2 \pmod{m}$ for some $x \in \mathbb{Z}$. If a satisfies $\gcd(a, m) = 1$ and is not a quadratic residue modulo m, it is called a (**quadratic**) **nonresidue**.*

It is clear that being a quadratic residue or not is a property of the congruence class of a. Often, but not always, we restrict our attention to the group $\mathbb{Z}_m^*$ that contains one representative from each congruence class in question. In this context, -1 always stands for the additive inverse of 1, i.e., for $m - 1$. Note that numbers a with $\gcd(a, m) > 1$ are considered neither quadratic residues nor nonresidues.

Example 6.1.2. For $m = 13$, the squares modulo 13 of $1, 2, \ldots, 12$ are $1, 4, 9, 3, 12, 10, 10, 12, 3, 9, 4, 1$, i.e., the quadratic residues are $1, 3, 4, 9, 10, 12$. For $m = 26$, the quadratic residues are $1, 3, 9, 17, 23, 25$; for $m = 27$, they are $1, 4, 7, 10, 13, 16, 19, 22, 25$.

We observe that in the case $m = 13$ there are 6 residues and 6 nonresidues. This behavior is typical for $m = p$ a prime number. If we square the numbers $1, \ldots, p - 1$, we obtain at most $\frac{1}{2}(p - 1)$ distinct values, since $x^2 \equiv (p - x)^2$ $\pmod{p}$. On the other hand, the squares of $1, \ldots, \frac{1}{2}(p - 1)$ are all distinct: if $x^2 \equiv y^2 \pmod{p}$ for $1 \leq x \leq y < \frac{1}{2}p$, then p divides $y^2 - x^2 = (x + y)(y - x)$;

M. Dietzfelbinger: Primality Testing in Polynomial Time, LNCS 3000, pp. 85-94, 2004.
© Springer-Verlag Berlin Heidelberg 2004

since $0 \leq y - x < x + y < p$, this can only be the case if $y = x$ or $y = p - x$. In other words, every $a \in \mathbb{Z}_p^*$ that is a quadratic residue modulo p has exactly two "square roots" modulo p, i.e., numbers $x \in \mathbb{Z}_p^*$ with $x^2 = a$ in $\mathbb{Z}_p^*$. — In the special case where p is a prime number it is not hard to find out whether $a \in \mathbb{Z}_p^*$ is a quadratic residue or not.

Lemma 6.1.3 (Euler's criterion). *If p is an odd prime number, then the set of quadratic residues is a subgroup of $\mathbb{Z}_p^*$ of size $(p-1)/2$. Moreover, for $a \in \mathbb{Z}_p^*$ (calculating in $\mathbb{Z}_p^*$), we have*

$$a^{(p-1)/2} = \begin{cases} 1, & \text{if } a \text{ is a quadratic residue modulo } p, \\ -1, & \text{if } a \text{ is a nonresidue modulo } p. \end{cases}$$

Proof. Since p is a prime number, the group $\mathbb{Z}_p^*$ is cyclic (Theorem 4.4.3). We calculate in this group. Let $g \in \mathbb{Z}_p^*$ be a primitive element — i.e., $\mathbb{Z}_p^* = \{1, g, g^2, g^3, \ldots, g^{p-2}\}$. Note that $g^{p-1} = 1$, and $g^{(p-1)/2} \neq 1$ is an element whose square is 1. Thus $g^{(p-1)/2} = -1$, since there are no nontrivial square roots of 1 in $\mathbb{Z}_p^*$. The $\frac{1}{2}(p-1)$ elements g^{2i}, $0 \leq i < p-1$, are the squares in this multiplicative group. (Note that if $\frac{1}{2}(p-1) \leq i < p-1$ then $(g^i)^2 = g^{2j}$ for $j = (i - (p-1)/2)$.) An element g^{2i} satisfies $(g^{2i})^{(p-1)/2} = g^{i(p-1)} = 1$. In contrast, an element g^{2i+1} satisfies $(g^{2i+1})^{(p-1)/2} = g^{i(p-1)} \cdot g^{(p-1)/2} = -1$. $\square$

For example, in $\mathbb{Z}_{13}$ the powers of the primitive element 2, obtained by repeated multiplication, are $\underline{1}, 2, \underline{4}, 8, \underline{3}, 6, \underline{12}, 11, \underline{9}, 5, \underline{10}, 7$, where the underlined elements are those with an even exponent. This representation of squares makes it easy to find a way to determine square roots modulo p for about half the primes p. Assume $p \geq 3$ is a prime number so that $p + 1$ is divisible by 4, like 7, 11, 19, 23, 31, and so on. Let g be a generating element of $\mathbb{Z}_p^*$, and let $a = g^{2i}$ be an arbitrary quadratic residue in $\mathbb{Z}_p^*$. Consider $x = a^{(p+1)/4} = g^{i(p-1)/2+i} = (-1)^i \cdot g^i$. Now $x^2 = (-1)^{2i} \cdot g^{2i} = 1 \cdot a = a$, which means that x is a square root of a. (The other one is $p - x$.) For finding x from a, we do not need to know g:

Lemma 6.1.4. *If p is an odd prime number with $p \equiv 3 \pmod 4$, then for each quadratic residue $a \in \mathbb{Z}_p^*$ the element $x = a^{(p+1)/4}$ satisfies $x^2 = a$.* $\square$

For example, consider $p = 11$. The number $a = 3$ satisfies $a^5 \bmod 11 = 1$, hence it is a quadratic residue. One square root of 3 is $3^{(11+1)/4} \bmod 11 = 3^3 \bmod 11 = 5$, the other one is $11 - 5 = 6$.

Finding square roots modulo p for prime numbers $p \equiv 1 \pmod 4$ is not quite as easy, but there are efficient (randomized) algorithms that perform this task. (See [16, p. 93ff.].)

There is an established notation for indicating whether a number $a \in \mathbb{Z}$ is a quadratic residue modulo an odd prime number p or not.

Definition 6.1.5. *For a prime number $p \geq 3$ and an integer a we let*

$$\left(\frac{a}{p}\right) = \begin{cases} 1, \text{ if } a \text{ is a quadratic residue modulo } p, \\ -1, \text{ if } a \text{ is a nonresidue modulo } p, \\ 0, \text{ if } a \text{ is a multiple of } p. \end{cases}$$

$\left(\frac{a}{p}\right)$ *is called the **Legendre symbol of** a **and** p.*

The Legendre symbol satisfies some simple rules, which are obvious from the definition and from Lemma 6.1.3.

Lemma 6.1.6. *For a prime number $p \geq 3$ we have:*

(a) $\left(\frac{a \cdot b}{p}\right) = \left(\frac{a}{p}\right) \cdot \left(\frac{b}{p}\right)$, *for all $a, b \in \mathbb{Z}$.*

(b) $\left(\frac{a \cdot b^2}{p}\right) = \left(\frac{a}{p}\right)$, *for $a, b \in \mathbb{Z}$, where $p \nmid b$.*

(c) $\left(\frac{a+cp}{p}\right) = \left(\frac{a}{p}\right)$, *for all integers a and c.*

   In particular, $\left(\frac{a}{p}\right) = \left(\frac{a \bmod p}{p}\right)$, *for all a.*

(d) $\left(\frac{-1}{p}\right) = (-1)^{(p-1)/2}$. □

Part (d) means that -1 (i.e., $p - 1$) is a quadratic residue modulo p if and only if $p \equiv 1 \pmod 4$. For example, 12 is a quadratic residue modulo 13, while 10 is a nonresidue modulo 11.

6.2 The Jacobi Symbol

Can we generalize the Legendre symbol to pairs a, n where $n \geq 3$ is an integer not assumed to be a prime number? There is an obvious way to do this: Assuming that $\gcd(a, n) = 0$, we could distinguish the cases where a is a quadratic residue modulo n and where a is a nonresidue. However, as it turns out, such a generalization would not have nice arithmetic and algorithmic properties. Instead, we use a technical definition.

Definition 6.2.1. *Let $n \geq 3$ be an odd integer with prime decomposition $n = p_1 \cdots p_r$. For integers a we let*

$$\left(\frac{a}{n}\right) = \left(\frac{a}{p_1}\right) \cdots \left(\frac{a}{p_r}\right).$$

$\left(\frac{a}{n}\right)$ *is called the **Jacobi symbol of** a **and** n.*

To get familiar with this definition, we consider some of its features. If n happens to be a prime number, then the Jacobi symbol $\left(\frac{a}{n}\right)$ is just the same as the Legendre symbol, so it is harmless that we use the same notation for both. If a and n have a common factor, then one of the p_i's divides a, hence $\left(\frac{a}{n}\right) = 0$ in this case. That means for example that $\left(\frac{9}{21}\right) = 0$ although 9 is a quadratic residue modulo 21. Interesting values occur only if a and n are relatively prime. It is important to note that even in this case $\left(\frac{a}{n}\right)$ does not indicate whether a is a quadratic residue modulo n or not. (For example, $\left(\frac{2}{15}\right) = \left(\frac{2}{3}\right) \cdot \left(\frac{2}{5}\right) = (-1)(-1) = 1$ while 2 is a quadratic nonresidue modulo 15.) Our definition will have the useful consequence that the Jacobi symbol is multiplicative in the upper and in the lower position, that square factors in the upper and in the lower position can be ignored, and that $\left(\frac{-1}{n}\right)$ can easily be evaluated.

Lemma 6.2.2. *If n, m range over odd integers ≥ 3, and a, b range over integers, then*

(a) $\left(\frac{a \cdot b}{n}\right) = \left(\frac{a}{n}\right) \cdot \left(\frac{b}{n}\right)$;

(b) $\left(\frac{a \cdot b^2}{n}\right) = \left(\frac{a}{n}\right)$, *if* $\gcd(b, n) = 1$;

(c) $\left(\frac{a}{n \cdot m}\right) = \left(\frac{a}{n}\right) \cdot \left(\frac{a}{m}\right)$;

(d) $\left(\frac{a}{n \cdot m^2}\right) = \left(\frac{a}{n}\right)$, *if* $\gcd(a, m) = 1$;

(e) $\left(\frac{a+cn}{n}\right) = \left(\frac{a}{n}\right)$, *for all integers c (in particular,* $\left(\frac{a}{n}\right) = \left(\frac{a \bmod n}{n}\right)$*)*;

(f) $\left(\frac{2^{2k} \cdot a}{n}\right) = \left(\frac{a}{n}\right)$ *and* $\left(\frac{2^{2k+1} \cdot a}{n}\right) = \left(\frac{2}{n}\right) \cdot \left(\frac{a}{n}\right)$, *for $k \geq 1$*;

(g) $\left(\frac{-1}{n}\right) = (-1)^{(n-1)/2}$;

(h) $\left(\frac{0}{n}\right) = 0$ *and* $\left(\frac{1}{n}\right) = 1$.

Proof. Parts (a) and (b) follow by applying Lemma 6.1.6(a) and (b) to the factors in Definition 6.2.1. — Part (c) is an immediate consequence of Definition 6.2.1. — Part (d), in turn, follows by noting that $\left(\frac{a}{m}\right) \in \{1, -1\}$ and applying (c) twice. — For part (e), we use that for every prime factor p of n we have that $\left(\frac{a+cn}{p}\right) = \left(\frac{a}{p}\right)$, by Lemma 6.1.6(c). — For part (f), we note that $\left(\frac{4}{n}\right) = \left(\frac{2}{n}\right)^2 = 1$, and then apply (b) repeatedly to eliminate factors of 4 in the upper position. — For part (g), we note the following. By definition, $\left(\frac{-1}{n}\right) = \left(\frac{-1}{p_1}\right) \cdots \left(\frac{-1}{p_r}\right)$, where $n = p_1 \cdots p_r$ is the prime number decomposition of n. We apply Lemma 6.1.6 to see that this product equals $(-1)^s$, where s is the number of indices i so that $p_i \equiv 3 \pmod 4$. On the other hand, we have

$$n \equiv (p_1 \bmod 4) \cdots (p_r \bmod 4) \equiv (-1)^s \pmod 4.$$

This means that $\frac{1}{2}(n-1)$ is odd if and only if s is odd, and hence $(-1)^s = (-1)^{(n-1)/2}$. — Part (h) is trivial. $\square$

6.3 The Law of Quadratic Reciprocity

In this section, we identify a method to efficiently calculate $\left(\frac{a}{n}\right)$. Most helpful for this is an amazing relationship that connects the values $\left(\frac{a}{n}\right)$ and $\left(\frac{n}{a}\right)$, for $a, n \geq 3$ odd.

To gather some intuition, we put up a table of some of the values $\left(\frac{a}{n}\right)$; see Table 6.1. We cover odd $a \geq 3$, but also include the special cases $a = -1$ (congruent modulo n to the odd number $2n - 1$) and $a = 2$ (congruent to $n + 2$).

n	a															
	3	5	7	9	11	13	15	17	19	21	23	25	27	29	−1	2
3	0	−	+	0	−	+	0	−	+	0	−	+	0	−	−	−
5	−	0	−	+	+	−	0	−	+	+	−	0	−	+	+	−
7	−	−	0	+	+	−	+	−	−	0	+	+	−	+	−	+
9	0	+	+	0	+	+	0	+	+	0	+	+	0	+	+	+
11	+	+	−	+	0	−	+	−	−	−	+	+	+	−	−	−
13	+	−	−	+	−	0	−	+	−	−	+	+	+	+	+	−
15	0	0	−	0	−	−	0	+	+	0	+	0	0	−	−	+
17	−	−	−	+	−	+	+	0	+	+	−	+	−	−	+	+
19	−	+	+	+	+	−	−	+	0	−	+	+	−	−	−	−
21	0	+	0	0	−	−	0	+	−	0	−	+	0	−	+	−
23	+	−	−	+	−	+	−	−	−	−	0	+	+	+	−	+
25	+	0	+	+	+	+	0	+	+	+	+	0	+	+	+	+
27	0	−	+	0	−	+	0	−	+	0	−	+	0	−	−	−
29	−	+	+	+	−	+	−	−	−	−	+	+	−	0	+	−

Table 6.1. Some values of the Jacobi symbol $\left(\frac{a}{n}\right)$. Instead of 1 we write +, instead of −1 we write −

Some features of the table are obvious: On the diagonal, there are 0's. In the column for $a = -1$, 1's and (-1)'s alternate as stated in Lemma 6.2.2(g). In the column for $a - 2$, a pattern shows up that suggests that $\left(\frac{2}{n}\right) = 1$ if $n \equiv 1$ or 7 (mod 8) and $\left(\frac{2}{n}\right) = -1$ if $n \equiv 3$ or 5 (mod 8). Columns and rows for square numbers like 9 and 25 show the expected pattern: $\left(\frac{a}{n^2}\right) = \left(\frac{n^2}{a}\right)$ equals 1 for all a with $\gcd(a, n) = 1$ and equals 0 otherwise. In row n, we have that $\left(\frac{2n+1}{n}\right) = \left(\frac{1}{n}\right) = 1$ and the entries repeat themselves periodically starting with $a = 2n + 3 \equiv 3$ (mod n). Further, since the Jacobi symbol is multiplicative both in the upper and in the lower position we can calculate the entries in row (column) $n_1 \cdot n_2$ from the entries in rows (columns) n_1 and n_2. For example, the entries in row (column) 27 are identical to those in row (column) 3.

One other pattern becomes apparent only after taking a closer look. If we compare row $n = 5$ with column $a = 5$, they turn out to be identical, at least within the small section covered by our table. Similar observations hold for $n = 13$ and $a = 13$, $n = 21$ and $a = 21$, or $n = 29$ and $a = 29$. For other corresponding pairs of rows and columns the situation is only a little more complicated. Row $n = 11$ and column $a = 11$ do not have the same entries, but if we compare carefully, we note that equal and opposite entries alternate: $\left(\frac{11}{3}\right) = -\left(\frac{3}{11}\right)$, $\left(\frac{11}{5}\right) = \left(\frac{5}{11}\right)$, $\left(\frac{11}{7}\right) = -\left(\frac{7}{11}\right)$, and so on. We make similar observations in rows (columns) for 3, for 19, and for 27. (The 0's strewn among these rows (columns) do not interrupt the pattern.) The rule that appears to govern the relationship between $\left(\frac{a}{n}\right)$ and $\left(\frac{n}{a}\right)$ is a famous fundamental theorem in number theory, called the *quadratic reciprocity law*. For the case of entries that are prime it was stated by Legendre; the first correct proof was given by Gauss. Since then, many proofs of this theorem have been given.

Theorem 6.3.1 (Quadratic Reciprocity Law). *If $m, n \geq 3$ are odd integers, then*

$$\left(\frac{m}{n}\right) = \begin{cases} \left(\frac{n}{m}\right), & \text{if } n \equiv 1 \text{ or } m \equiv 1 \pmod 4, \\ -\left(\frac{n}{m}\right), & \text{if } n \equiv 3 \text{ and } m \equiv 3 \pmod 4. \end{cases}$$

Proposition 6.3.2. *If $n \geq 3$ is an odd integer, then*

$$\left(\frac{2}{n}\right) = \begin{cases} 1, & \text{if } n \equiv 1 \text{ or } n \equiv 7 \pmod 8, \\ -1, & \text{if } n \equiv 3 \text{ or } n \equiv 5 \pmod 8. \end{cases}$$

The proofs of these two theorems may be found in Sect. A.3 in the appendix. Here, we notice how these laws give rise to an extremely efficient procedure for evaluating the Jacobi symbol $\left(\frac{a}{n}\right)$ for any odd integer $n \geq 3$ and any integer a. The idea is easiest formulated recursively: Let an arbitrary integer a and an odd number $n \geq 3$ be given.

(1) If a is not in the interval $\{1, \ldots, n-1\}$, the result is $\left(\frac{a \bmod n}{n}\right)$.
(2) If $a = 0$, the result is 0.
(3) If $a = 1$, the result is 1.
(4) If $4 \mid a$, the result is $\left(\frac{a/4}{n}\right)$.
(5) If $2 \mid a$, the result is $\left(\frac{a/2}{n}\right)$ if $n \bmod 8 \in \{1, 7\}$ and $-\left(\frac{a/2}{n}\right)$ if $n \bmod 8 \in \{3, 5\}$.
(6) If ($a > 1$ and) $a \equiv 1$ or $n \equiv 1 \pmod 4$, the result is $\left(\frac{n \bmod a}{a}\right)$.
(7) If $a \equiv 3$ and $n \equiv 3 \pmod 4$, the result is $-\left(\frac{n \bmod a}{a}\right)$.

These rules make it easy to calculate $\left(\frac{a}{n}\right)$ by hand for a and n that are not too large. For example, assume we wish to find the value $\left(\frac{773}{1373}\right)$. (Both 773 and 1373 are prime numbers, so in fact we are asking for the Legendre symbol $\left(\frac{773}{1373}\right)$.) Applying our rules, we obtain:

$$\left(\frac{773}{1373}\right) \overset{(6)}{=} \left(\frac{600}{773}\right) \overset{(4)}{=} \left(\frac{150}{173}\right) \overset{(5)}{=} -\left(\frac{75}{173}\right)$$
$$\overset{(6)}{=} -\left(\frac{23}{75}\right) \overset{(7)}{=} \left(\frac{6}{23}\right) \overset{(5)}{=} \left(\frac{3}{23}\right) \overset{(6)}{=} -\left(\frac{2}{3}\right) \overset{(5)}{=} \left(\frac{1}{3}\right) \overset{(3)}{=} 1.$$

Thus, 773 is a quadratic residue modulo 1373 (which we could have found out also by calculating $773^{686} \bmod 1373 = 1$).

For an implementation, we prefer an iterative procedure, as suggested by the example. It is sufficient to systematically apply rules (1)–(7) and to accumulate the factors (-1) introduced by applying the rules in a variable s ("sign"). After some streamlining, one arrives at the following algorithm.

Algorithm 6.3.3 (Jacobi Symbol)

INPUT: Integer a, odd integer $n \geq 3$.
METHOD:

```
0     b, c, s: integer;
1     b ← a mod n; c ← n;
2     s ← 1;
3     while b ≥ 2 repeat
4          while 4 | b repeat b ← b/4;
5          if 2 | b then
6               if c mod 8 ∈ {3,5} then s ← (−s);
7               b ← b/2;
8          if b = 1 then break;
9          if b mod 4 = c mod 4 = 3 then s ← (−s);
10         (b, c) ← (c mod b, b);
11    return s · b;
```

In order to understand what this algorithm does one should imagine that in b and c two coefficients b and c are stored, which indicate that the Jacobi symbol $\left(\frac{b}{c}\right)$ is still to be evaluated. The big **while** loop (lines 3–10) is iterated until the upper component b has reached a value in $\{0, 1\}$, at which point we have $b = \left(\frac{b}{c}\right)$. The variable s contains a number $s \in \{-1, 1\}$, which accumulates the sign changes caused by previous iterations. In one iteration of the while loop first the upper component b is reduced to its largest odd factor, while s is switched if appropriate. Then (lines 9–10) the quadratic reciprocity law is applied once, again changing s if needed. Note that no extra evaluation of the greatest common divisor of a and n is needed; if $\gcd(a, n) > 1$, this is detected by b becoming 0 at some point, while $c \geq 3$. (In fact it is quite easy to see that we always have $\gcd(b, c) = \gcd(a, n)$: Since c is odd, dividing b by 4 and 2 does not change the greatest common divisor; the operation in lines 9–10 is covered by Proposition 3.1.10.)

Proposition 6.3.4. *Algorithm 6.3.3 outputs the value $\left(\frac{a}{n}\right)$. The number of iterations of the* **while** *loop in lines 3–10 is $O(\log n)$.*

Proof. We claim that the contents b, c, and s of b, c, and s satisfy the invariant

$$s \cdot \left(\frac{b}{c}\right) = \left(\frac{a}{n}\right), \tag{6.3.1}$$

whenever line 2, 4, 7, or 10 has been carried out. This is trivially true at the beginning (after line 2). Dividing b by 4 in line 4 does not change the value $\left(\frac{b}{c}\right)$, see Lemma 6.2.2(f). Dividing b by 2 in line 7 is compensated by multiplying s by $\left(\frac{2}{c}\right)$ in line 6; see Proposition 6.3.2. Note that after line 7 has been carried out, b is an odd number. The test in line 8 makes sure that the loop is left if b has become 1; when we reach line 9, b is guaranteed to be an odd number ≥ 3. In line 10 $\left(\frac{b}{c}\right)$ is replaced by $\left(\frac{c \bmod b}{b}\right) = \left(\frac{c}{b}\right)$. If necessary, s is updated in line 9. Lemma 6.2.2(e) and Theorem 6.3.1 guarantee that (6.3.1) continues to hold.

Next, we consider the number of iterations of the big while loop (lines 3–10). Here, the analysis is similar to the analysis of the number of rounds in the Euclidean Algorithm 3.2.1. One shows that if the contents of b and c at the times when line 3 is executed are $(b_0, c_0), (b_1, c_1), \ldots$, then $c_{t+2} \leq \frac{1}{2} c_t$ for $t = 0, 1, 2, \ldots$. Since $c_0 = n$, the loop cannot be carried out more often than $2\|n\| = O(\log n)$.

Summing up, after a logarithmic number of iterations the while loop stops and line 11 is reached. At this point, $b \in \{0, 1\}$, and hence $\left(\frac{b}{c}\right) = b$. By (6.3.1) we see that the correct value $s \cdot b = s \cdot \left(\frac{b}{c}\right) = \left(\frac{a}{n}\right)$ is returned. $\square$

One last comment on Algorithm 6.3.3 concerns the complexity in terms of bit operations. One should note that dividing a number b given in binary by 2 or by 4 amounts to dropping one or two trailing 0's. Determining the remainder of b and c modulo 4 or modulo 8 amounts to looking at the last two or three bits of c. So the only costly operations in the algorithm are the divisions with remainder in lines 1 and 10. Thus, from the point of view of computational efficiency Algorithm 6.3.3 is comparable to the simple Euclidean Algorithm (see Lemma 3.2.3) — an amazing consequence of the Quadratic Reciprocity Law!

6.4 Primality Testing by Quadratic Residues

In order to take full advantage of this section, the reader is advised to recall Sect. 5.1 on the Fermat test. — Let $p \geq 3$ be a prime number, and let $1 \leq a < p$. In the last two sections we have developed two ways for finding out whether a is a quadratic residue modulo p: we could evaluate $a^{(p-1)/2} \bmod p$ by fast exponentiation, or we could evaluate the Legendre symbol $\left(\frac{a}{p}\right)$ as a Jacobi symbol, using Algorithm 6.3.3. In both cases, the result is 1 if a is a quadratic residue modulo p and $-1 \equiv p - 1$ otherwise. We summarize:

Lemma 6.4.1. *If p is an odd prime number, then*

$$a^{(p-1)/2} \cdot \left(\frac{a}{p}\right) \bmod p = 1, \text{ for all } a \in \{1, \ldots, p-1\}.$$

$\square$

Turning the lemma around, we note that if $n \geq 3$ is odd and $a \in \{2, \ldots, n-1\}$ satisfies $a^{(n-1)/2} \cdot \left(\frac{a}{n}\right) \bmod n \neq 1$, then n cannot be a prime number. Recall from Sect. 5.1 that we called an element a, $1 \leq a < n$, an F-witness for an odd composite number n if $a^{n-1} \bmod n \neq 1$, and an F-liar for n otherwise. In reference to the Euler criterion from Lemma 6.1.3 we now define:

Definition 6.4.2. *Let n be an odd composite number. A number a, $1 \leq a < n$, is called an **E-witness** for n if $a^{(n-1)/2} \cdot \left(\frac{a}{n}\right) \bmod n \neq 1$. It is called an **E-liar** otherwise.*

Example 6.4.3. Consider the composite number $n = 325$. For $a = 15$, we have $\gcd(15, 325) = 5$, hence $\left(\frac{15}{325}\right) = 0$, and 15 is an E-witness. For $a = 2$, we have $2^{162} \bmod 325 = 129$, so 2 is an E-witness as well. For $a = 7$, we have $7^{162} \bmod 325 = 324$ and $\left(\frac{7}{325}\right) = -1$; this means that 7 is an E-liar for 325.

Lemma 6.4.4. *Let $n \geq 3$ be an odd composite number. Then every E-liar for n also is an F-liar for n.*

Proof. If a is an E-liar, then $1 = a^{(n-1)/2} \cdot \left(\frac{a}{n}\right) \bmod n$, hence $\left(\frac{a}{n}\right) \in \{1, -1\}$ and $1 = (a^{(n-1)/2} \cdot \left(\frac{a}{n}\right))^2 \bmod n = a^{n-1} \bmod n$. So, a is an F-liar. $\square$

In the following we show that for all odd composite numbers $n \geq 3$ the E-liars for n can make up at most half of the elements of $\mathbb{Z}_n^*$. This is the basis for a randomized primality test that can be used as an alternative to the Miller-Rabin test.

Lemma 6.4.5. *Let $n \geq 3$ be an odd composite number. Then the set $\{a \mid a$ is an E-liar for $n\}$ is a proper subgroup of $\mathbb{Z}_n^*$.*

Proof. We know from Lemma 5.1.3(a) that the set of F-liars for n is a subset of $\mathbb{Z}_n^*$. By Lemma 6.4.4 it follows that all E-liars are in $\mathbb{Z}_n^*$. Now we use the subgroup criterion, Lemma 4.1.6: (i) Clearly, 1 is an E-liar. (ii) Assume $a, b \in \mathbb{Z}_n^*$ are E-liars. Then

$$(a \cdot b)^{(n-1)/2} \cdot \left(\frac{a \cdot b}{n}\right) \bmod n$$

$$= (a^{(n-1)/2} \cdot \left(\frac{a}{n}\right) \bmod n) \cdot (b^{(n-1)/2} \cdot \left(\frac{b}{n}\right) \bmod n) = 1 \cdot 1 = 1,$$

using multiplicativity of the Jacobi symbol (Lemma 6.2.2(a)).

It remains to show that there is at least one E-witness in $\mathbb{Z}_n^*$. We consider two cases.

Case 1: n is divisible by p^2, for some prime number $p \geq 3$. — In the proof of Lemma 5.1.8 we have seen how to construct an F-witness a in $\mathbb{Z}_n^*$ in this case. By Lemma 6.4.4, a is also an E-witness.

Case 2: n is a product of several distinct prime numbers. — Then we may write $n = p \cdot m$ for p an odd prime number and $m \geq 3$ odd with $p \nmid m$. Let $b \in \mathbb{Z}_p^*$ be some quadratic nonresidue modulo p. This means that $\left(\frac{b}{p}\right) = -1$. Applying the Chinese Remainder Theorem 3.4.1, we see that there is some a, $1 \leq a < n$, with

$a \equiv b \pmod{p}$ and

$a \equiv 1 \pmod{m}$.

Claim: $a \in \mathbb{Z}_n^*$ and a is an E-witness.
Proof of Claim: Clearly, $p \nmid a$ and $\gcd(a, m) = 1$, so a is in $\mathbb{Z}_n^*$. Next we note that

$$\left(\frac{a}{n}\right) = \left(\frac{a}{p}\right) \cdot \left(\frac{a}{m}\right) = \left(\frac{b}{p}\right) \cdot \left(\frac{1}{m}\right) = (-1) \cdot 1 = -1. \tag{6.4.2}$$

Now if a were an E-liar, we would have, in view of (6.4.2), that $a^{(n-1)/2} \equiv -1 \pmod{n}$. Since m is a divisor of n, this would entail

$a^{(n-1)/2} \equiv -1 \pmod{m}$,

which contradicts the fact that $a \equiv 1 \pmod{m}$. So a must be an E-witness for n. $\qquad \square$

In combination with Proposition 4.1.9, Lemma 6.4.5 entails that the number of E-liars for n is a proper divisor of $|\mathbb{Z}_n^*| = \varphi(n)$, which means that at least half of the elements of $\mathbb{Z}_n^*$ are E-witnesses. We formulate the resulting primality test.

Algorithm 6.4.6 (Solovay-Strassen Test)

INPUT: Odd integer $n \geq 3$.
METHOD:
1   Let a be randomly chosen from $\{2, \ldots, n-2\}$;
2   **if** $a^{(n-1)/2} \cdot \left(\frac{a}{n}\right) \bmod n \neq 1$
3   **then return** 1;
4   **else return** 0;

It is understood that for calculating the Jacobi symbol $\left(\frac{a}{n}\right)$ in line 2 we use Algorithm 6.3.3, and for calculating the power $a^{(n-1)/2} \bmod n$ we use Algorithm 2.3.3.

Proposition 6.4.7. *Algorithm 6.4.6, when applied to an input number n, needs $O(\log n)$ arithmetic operations on numbers smaller than n^2, which amounts to $O((\log n)^3)$ bit operations (naive methods) resp. $O^{\sim}((\log n)^2)$ bit operations (best methods). If n is a prime number, the output is 0, if n is composite, the probability that output 0 is given is smaller than $\frac{1}{2}$.*

Proof. The time and bit operation bounds are those of Algorithm 2.3.3, which are at least as big as those of Algorithm 6.3.3. Lemma 6.4.1 tells us that if n is a prime number the algorithm outputs 0. If n is composite, the algorithm outputs 0 if the value a chosen at random happens to be an E-liar. By Lemma 6.4.5 we know that the set of E-liars is a proper subgroup of $\mathbb{Z}_n^*$, and hence comprises no more than $\frac{1}{2}\varphi(n)$ elements. Exactly as in the case of the Fermat test (inequality (5.1.1)) we see that the probability that an element a randomly chosen from $\{2, \ldots, n-2\}$ is an E-liar is smaller than $\frac{1}{2}$. $\qquad \square$

7. More Algebra: Polynomials and Fields

In preparation for the correctness proof of the deterministic primality test in Chap. 8, in this chapter we develop a part of the theory of polynomials over rings and fields, and study how fields arise from polynomial rings by calculating modulo an irreducible polynomial.

7.1 Polynomials over Rings

The reader should recall the definition of (commutative) rings (with 1) and fields from Sect. 4.3, as well as the two central examples: the structure $(\mathbb{Z}_m, +_m, \cdot_m, 0, 1)$ is a finite ring for each $m \geq 2$, and it is a field if and only if m is a prime number.

In calculus, an important object of studies are "polynomial functions" like

$$f(x) = 10x^3 + 4.4x^2 - 2,$$

interpreted over the field of real numbers (or the field of complex numbers). We obtain such a polynomial by multiplying some powers $x^d, \ldots, x = x^1, 1 = x^0$ of a "variable" x with real coefficients $a_d, \ldots, a_2, a_1, a_0$ and adding the resulting terms; in formulas:

$$f(x) = a_d \cdot x^d + \cdots + a_2 \cdot x^2 + a_1 \cdot x + a_0. \tag{7.1.1}$$

If we now imagine that x ranges over all real numbers as arguments, and consider the values obtained by evaluating the polynomial f for these arguments, we obtain the function

$$\mathbb{R} \ni x \mapsto f(x) \in \mathbb{R}.$$

Interpreted in this way, the real polynomials just form a certain subset of all functions from $\mathbb{R}$ to $\mathbb{R}$, namely those functions that may be represented by an expression such as (7.1.1). Of course, we can consider the classes of functions defined by polynomials over arbitrary fields and rings, not only over the real numbers.

M. Dietzfelbinger: Primality Testing in Polynomial Time, LNCS 3000, pp. 95-114, 2004.
© Springer-Verlag Berlin Heidelberg 2004

In this book, as is standard in algebra, polynomials over a ring R will be used in a somewhat more abstract way. In particular, the "variable" X (as is quite common in this context, we use an uppercase letter to denote the variable) used in writing the polynomials is not immediately meant to be replaced by some element of R. Rather, using this variable, a polynomial is declared to be just a "formal expression"

$$a_d X^d + \cdots + a_2 X^2 + a_1 X + a_0,$$

where the "coefficients" $a_d, \ldots, a_2, a_1, a_0$ are taken from R. The resulting expressions are then treated as objects in their own right. They are given an arithmetic structure by mimicking the rules for manipulating real polynomials: polynomials f and g are added by adding the coefficients of identical powers of X in f and g, and f and g are multiplied by multiplying every term $a_i X^i$ by every term $b_j X^j$, transforming $(a_i X^i) \cdot (b_j X^j)$ into $(a_i \cdot b_j) X^{i+j}$, and adding up the coefficients associated with the same power of X. (The coefficients a_0 and a_1 are treated as if they were associated with X^0 and X^1, respectively.)

Example 7.1.1. Starting from the ring $\mathbb{Z}_{12}$, with $+_{12}$ and $\cdot_{12}$ denoting addition and multiplication modulo 12, consider the two polynomials $f = 3X^4 + 5X^2 + X$ and $g = 8X^2 + X + 3$. (We follow the standard convention that coefficients that are 1 and terms $0X^i$ are omitted.) Then

$$f + g = 3X^4 + (5 +_{12} 8)X^2 + (1 +_{12} 1)X + (0 +_{12} 3) = 3X^4 + X^2 + 2X + 3$$

and

$$f \cdot g = 3X^5 + X^4 + X^3 + 4X^2 + 3X,$$

since $3 \cdot_{12} 8 = 0$, $3 \cdot_{12} 1 + 0 \cdot 0 = 3$, $3 \cdot_{12} 3 +_{12} 0 \cdot_{12} 1 +_{12} 5 \cdot_{12} 8 = 1$, and so on.

Writing polynomials as formal sums of terms $a_i X^i$ has the advantage of delivering a picture close to our intuition from real polynomials, but it has notational disadvantages, e.g., we might ask ourselves if there is a difference between $3X^3 + 0X^2 + 2X$, $0X^4 + 3X^3 + 0X^2 + 2X + 0$, and $2X + 3X^3$, or if these seemingly different formal expressions should just be regarded as different names for the same "object". Also, the question may be asked what kind of object X is and if the "+"-signs have any "meaning". The following (standard) formal definition solves all these questions in an elegant way, by omitting the X's and +'s in the definition of polynomials altogether.

We note that the only essential information needed to do calculations with a real polynomial is the sequence of its coefficients. Consequently, we represent a polynomial over a ring R by a formally infinite coefficient sequence $(a_0, a_1, a_2, \ldots)$, in which only finitely many nonzero entries appear. (Note the reversal of the order in the notation.)

Definition 7.1.2. *Let $(R, +, \cdot, 0, 1)$ be a ring. The set $R[X]$ is defined to be the set of all (formally infinite) sequences*

$$(a_0, a_1, \ldots), \quad a_0, a_1, \ldots \in R, \quad \text{where all but finitely many } a_i \text{ are } 0.$$

*The elements of $R[X]$ are called the **polynomials over** R (or, more precisely: the polynomials over R **in one variable**). Polynomials are denoted by $f, g, h, \ldots$.*

For convenience, we allow ourselves to write $(a_0, a_1, \ldots, a_d)$ for the sequence $(a_0, a_1, \ldots)$, if $a_i = 0$ for all $i > d \geq 0$, without implying that a_d should be nonzero. For example, the sequences $(0, 1, 5, 0, 3, 0, 0, \ldots)$, $(0, 1, 5, 0, 3)$, and $(0, 1, 5, 0, 3, 0)$ all denote the same polynomial.

On $R[X]$ we define two operations, for convenience denoted by $+$ and $\cdot$ again. Addition is carried out componentwise; multiplication is defined in the way suggested by the informal description given above.

Definition 7.1.3. *Let $(R, +, \cdot, 0, 1)$ be a ring. For $f = (a_0, a_1, \ldots)$ and $g = (b_0, b_1, \ldots)$ in $R[X]$ let*

(a) $f + g = (a_0 + b_0, a_1 + b_1, \ldots)$, *and*
(b) $f \cdot g = (c_0, c_1, \ldots)$, *where $c_i = (a_0 \cdot b_i) + (a_1 \cdot b_{i-1}) + \cdots + (a_i \cdot b_0)$, for $i = 0, 1, \ldots$. (Only finitely many of the c_i can be nonzero.)*

Example 7.1.4. (a) In the ring $\mathbb{Z}_{12}[X]$, the two polynomials f and g from Example 7.1.1 would be written $f = (0, 1, 5, 0, 3)$ and $g = (3, 1, 8)$ (omitting trailing zeroes). Further, $f +_{12} g = (3, 2, 1, 0, 3)$ and $f \cdot_{12} g = (0, 3, 4, 1, 1, 3)$, as is easily checked. The two polynomials $(0, 3, 6)$ and $(8, 4)$ have product $(0, 0, \ldots)$, or (0), the zero polynomial.
(b) In $\mathbb{Z}[X]$, the polynomials $f = (0, 1, -5, 0, 10)$ and $g = (0, -1, 5, 0, -10)$ satisfy $f + g = (0)$, and $f \cdot g = (0, 0, -1, 10, -25, -20, 100, 0, -100)$, as the reader may want to check.

Proposition 7.1.5. *If $(R, +, \cdot, 0, 1)$ is a ring, then $(R[X], +, \cdot, (0), (1))$ is also a ring.*

Proof. We must check the conditions (i), (ii), and (iii) in Definition 4.3.2. For (i) and (ii), it is a matter of routine to see that $(R[X], +, (0))$ is an abelian group, and that $(R[X], \cdot, (1))$ is a commutative monoid. Finally, for (iii), it takes a little patience, but it just requires a straightforward calculation on the basis of the definitions to check that $(f + g) \cdot h = (f \cdot h) + (g \cdot h)$, for $f, g, h \in R[X]$.   □

It is obvious that $R[X]$ contains a subset that is isomorphic to R, namely the set of elements $(a) = (a, 0, 0, \cdots)$, $a \in R$. Addition and multiplication of these elements has an effect only on the first component, and works exactly as in R. Thus, we may identify element $a \in R$ with $(a) \in R[X]$ and regard R as a "subring" of $R[X]$. (The precise definition of a subring is given as

Definition 7.1.10 below.) In particular, the zero polynomial $(0) = (0, 0, \ldots)$ is denoted by 0, and the polynomial $(1) = (1, 0, 0, \ldots)$ is denoted by 1. Polynomials that are not elements of R are often called **nonconstant**.

In a move that might look a little arbitrary at first, we pick out the special element $(0, 1) = (0, 1, 0, 0, \ldots)$ of $R[X]$ and name it X. It is easily checked that the powers of X satisfy $X^0 = 1 = (1), X^1 = X = (0, 1), X^2 = (0, 0, 1), X^3 = (0, 0, 0, 1), \ldots$, and in general

$$X^i = (0, \ldots, 0, 1, 0, 0, \ldots), \text{ for } i \geq 0,$$

where the "1" appears in the $(i+1)$st position of the sequence. Together with the rules for addition and multiplication in $R[X]$ this shows that an arbitrary polynomial $f = (a_0, a_1, \ldots, a_d)$ can be written as

$$f = a_0 \cdot X^0 + a_1 \cdot X^1 + a_2 \cdot X^2 + \cdots + a_d \cdot X^d = a_0 + a_1 X + a_2 X^2 + \cdots + a_d X^d,$$

but now not as a "formal expression", but as a legitimate expression in the ring $R[X]$. It is very important to note that there is only one way to write a polynomial in this manner: if $a_0 + a_1 X + a_2 X^2 + \cdots + a_d X^d = b_0 + b_1 X + \cdots + b_{d'} X^{d'}$, for $d \leq d'$, say, then $(a_0, \ldots, a_d, 0, 0, \ldots) = (b_0, \ldots, b_{d'}, 0, 0, \ldots)$, which means $a_i = b_i$ for $i \leq \min\{d, d'\}$ and $a_i = 0$ for $d < i \leq d'$. This almost trivial observation is used in a useful technique called "comparison of coefficients".

As is common, the highest power of X that appears in a polynomial f with a nonzero coefficient is called the "degree" of f. More formally:

Definition 7.1.6. *For $f = (a_0, a_1, \ldots) \in R[X]$ we let*

$$\deg(f) = \begin{cases} -\infty & \text{, if } f = 0, \\ \max\{i \mid a_i \neq 0\} & \text{, if } f \neq 0, \end{cases}$$

*and call $\deg(f)$ the **degree** of f.*

If $f \neq 0$, and $d = \deg(f)$, we can write $f = (a_0, \ldots, a_d) = a_d X^d + \cdots + a_1 X + a_0$ with $a_d \neq 0$. In this case, a_d is called the **leading coefficient** of f. Extending standard arithmetic rules, we define $(-\infty) + d = d + (-\infty) = -\infty$ for $d \in \mathbb{N} \cup \{-\infty\}$ and declare that $-\infty < d$ for all $d \in \mathbb{N}$. Then we have the following elementary rules for the combination of degrees and the ring operations in $R[X]$.

Proposition 7.1.7. *For $f, g \in R[X]$ we have*

$$\deg(f + g) \leq \max\{\deg(f), \deg(g)\} \text{ and } \deg(f \cdot g) \leq \deg(f) + \deg(g).$$

Proof. The case of addition is obvious. In the case of multiplication note that by Definition 7.1.3(b) in $(c_0, c_1, \ldots) = (a_0, \ldots, a_d) \cdot (b_0, \ldots, b_{d'})$ all c_i with $i > d + d'$ must be 0. □

Remark 7.1.8. It is easy to implement a representation of polynomials: a polynomial $f = (a_0, \ldots, a_d)$ is represented as a vector of $d+1$ elements from R. (Leading zeroes may be inserted if needed.) The implementation of addition and subtraction of polynomials is then straightforward; the cost is $d+1$ ring operations if both arguments have degree at most d. We may multiply two polynomials f and g at the cost of $\deg(f) \cdot \deg(g)$ multiplications and additions in R. Note, however, that there are faster methods for multiplying polynomials over rings R, which require only $O(d(\log d)(\log \log d))$ multiplications of elements of R for multiplying two degree-d polynomials ([41, Sect. 8.3]).

The reader should note that it is possible that $\deg(f + g) < \max\{\deg(f), \deg(g)\}$ (if $\deg(f) = \deg(g)$ and the leading coefficients add to 0) and that $\deg(f \cdot g) < \deg(f) + \deg(g)$. The latter situation occurs if the leading coefficients of f and g have product 0 in R. For this to be possible R must contain zero divisors, e.g., $R = \mathbb{Z}_m$ where m is a composite number. We are particularly interested in those situations in which this cannot happen.

A polynomial $f \neq 0$ is called **monic** if its leading coefficient is 1.

Lemma 7.1.9. *Let $f, g \in R[X]$. Then we have:*

(a) *if f is monic, then $\deg(f \cdot g) = \deg(f) + \deg(g)$;*
(b) *if a is a unit in R, then $\deg(a \cdot g) = \deg(g)$;*
(c) *if $f \neq 0$ and has a unit as leading coefficient, then $\deg(f \cdot g) = \deg(f) + \deg(g)$.*

Proof. (a) Let $d \in \mathbb{N}$, $f = (a_0, a_1, \ldots, a_d)$ with $a_d = 1$. For $g = 0$ the claim is obviously true; hence assume $g = (b_0, \ldots, b_{d'})$ with $b_{d'} \neq 0$. Write $f \cdot g = (c_0, \ldots, c_{d+d'})$, for some $c_0, \ldots, c_{d+d'} \in R$. Clearly, $c_{d+d'} = 1 \cdot b_{d'} = b_{d'} \neq 0$, hence $\deg(f \cdot h) = d + d'$.
(b) The case $g = 0$ is trivial. Thus, consider $g = (b_0, \ldots, b_{d'})$ with $b_{d'} \neq 0$. Then $a \cdot g = (a \cdot b_0, \ldots, a \cdot b_{d'})$. By the remarks after Definition 4.3.4, $a \cdot b_{d'} \neq 0$, since the unit a cannot divide 0.
(c) Let $f = (a_0, a_1, \ldots, a_d)$ with $u \cdot a_d = 1$. Then $f_1 = u \cdot f$ is a monic polynomial with $\deg(f_1) = \deg(f)$, and by (i) and (ii) we get $\deg(f \cdot g) = \deg(a_d \cdot (f_1 \cdot g)) = \deg(f) + \deg(g)$. $\qquad \square$

For example, in the ring $\mathbb{Z}_{12}[X]$, we know without calculation that $\deg((5X^4 + 6) \cdot (2X^2 + 4X + 2)) = 6$, because 5 is a unit in this ring. In contrast, $\deg((6X^4 + 3) \cdot (4X^2 + 8X + 4)) = -\infty$ in the same polynomial ring.

As the last part of our general considerations for polynomials, we define what it means to "*substitute*" a value for X in a polynomial $f \in R[X]$. (Even this operation of substitution does not make X anything else but an ordinary element of the ring $R[X]$.) We need the notion of a ring being a substructure of another ring.

Definition 7.1.10. *Let $(S, +, \cdot, 0, 1)$ be a ring. A set $R \subseteq S$ is a **subring** of S if R contains 0 and 1, is closed under $+$ and $\cdot$, and $(R, +, \cdot, 0, 1)$ is a ring.*

Example 7.1.11. (a) The ring $\mathbb{Z}$ is a subring of the ring $\mathbb{Q}$; in turn, $\mathbb{Q}$ is a
  subring of the ring $\mathbb{R}$; finally, $\mathbb{R}$ is a subring of $\mathbb{C}$, the field of complex
  numbers.
(b) If R is any ring, it is a subring of $R[X]$, as discussed above.
(c) Although $\{0, 1, \ldots, m-1\} \subseteq \mathbb{Z}$, the ring $\mathbb{Z}_m$ is *not* a subring of $\mathbb{Z}$, since
  different operations are used in the two structures.

Definition 7.1.12. *Assume that R is a subring of the ring S, and that $s \in S$.
For $f = (a_0, a_1, \ldots, a_d) = a_d X^d + \cdots + a_1 X + a_0 \in R[X]$ define*

$$f(s) = a_d \cdot s^d + \cdots + a_1 \cdot s + a_0.$$

*We say that $f(s)$ results from **substituting** s in f.*

Proposition 7.1.13. *In the situation of the preceding definition we have:*

(a) $f(s) = a$ *if* $f = a \in R$,
(b) $f(s) = s$ *if* $f = X$, *and*
(c) $(f + g)(s) = f(s) + g(s)$ *and* $(f \cdot g)(s) = f(s) \cdot g(s)$, *for all* $f, g \in R[X]$.

*(In brief, the mapping $f \mapsto f(s)$ is a "ring homomorphism" from $R[X]$ to S
that leaves all elements of R fixed and maps X to s.)*

Proof. (a) and (b) are immediate from the definition of $f(b)$, if we apply
it to $f = a$ and $f = X$, respectively. (c) is proved by a straightforward
calculation, which we carry out for the case of multiplication. Consider $f =
a_d X^d + \cdots + a_1 X + a_0$ and $g = b_d X^d + \cdots + b_1 X + b_0$. Then by the definition
of multiplication in $R[X]$ we have $f \cdot g = c_{2d} X^{2d} + \cdots + c_1 X + c_0$, where
$c_k = \sum_{0 \le i,j \le d, i+j=k} (a_i + b_j)$, for $0 \le k \le 2d$. From this we get by substituting

$$(f \cdot g)(s) = c_{2d} \cdot s^{2d} + \cdots + c_1 \cdot s + c_0.$$

On the other hand, $f(s) = a_d \cdot s^d + \cdots + a_1 \cdot s + a_0$ and $g(s) = b_d \cdot s^d + \cdots +
b_1 \cdot s + b_0$, hence, by multiplication in S,

$$f(s) \cdot g(s) = c_{2d} \cdot s^{2d} + \cdots + c_1 \cdot s + c_0.$$

(It should be clear from this argument that the main reason to define multiplication of polynomials in the way done in Definition 7.1.3(b) was to make
this proposition true.)   □
  We mention two examples of such substitutions.

Example 7.1.14. (a) Consider the ring $\mathbb{R}$ with its subring $\mathbb{Z}$. If we substitute
  the element $s = \frac{1}{2}(\sqrt{5} + 1) \approx 1.61803$ in the integer polynomial $f =
  (-1, -1, 1, 0, 0, \ldots)$ or $f = X^2 - X - 1$, we obtain the value $f(s) = s^2 -
  s - 1 = \frac{1}{4}(6 + 2\sqrt{5}) - \frac{1}{2}(\sqrt{5} + 1) - 1 = 0$. We say that s is a "root" of
  f. Note that to find a root of f, we have to go beyond the ring $\mathbb{Z}$. More
  generally, we may substitute arbitrary real or complex values in arbitrary
  polynomials with integer coefficients.

(b) If R is an arbitrary ring, we may regard R as a subring of $S = R[X]$, and substitute elements of $R[X]$ in f. For example, we may substitute X^2 in $f = (a_0, a_1, a_2, \ldots)$ to obtain $f(X^2) = (a_0, 0, a_1, 0, a_2, 0, \ldots)$. Likewise, we can substitute other powers of X or, more generally, arbitrary polynomials g. For example, if $f = X^2$ then $f(g) = g^2$, if $f = X^d$ then $f(g) = g^d$. In particular, for $f = X$ we obtain $f(g) = g$. The other way round, it is important to notice what we get if we let the element $X = (0,1) \in R[X]$ itself play the role of the element to be substituted: for arbitrary $f \in R[X]$ we have $f(X) = f$. Thus $f(X)$ is just a wordy way of writing f, which we will use when it is convenient.

For later use, we note a relation between powers of polynomials and polynomials of powers of X, over $\mathbb{Z}_p$ for p a prime number. Before we give the statement and the proof, we look at an example. Let $f = 2X^3 + X^2 + 2$, and $p = 3$. Then, as a slightly tedious calculation over $\mathbb{Z}_3$ shows, we have $f^3 = (2X^3 + X^2 + 2)(2X^3 + X^2 + 2)(2X^3 + X^2 + 2) = 2X^9 + X^6 + 2 - f(X^3)$. The reason for this result is the following general fact.

Proposition 7.1.15. *Let p be a prime number. Then*

(a) $(f + g)^p = f^p + g^p$ *and* $(f \cdot g)^p = f^p \cdot g^p$ *, for* $f, g \in \mathbb{Z}_p[X]$ *;*
(b) *for* $f \in \mathbb{Z}_p[X]$ *we have* $f^p = f(X^p)$*, and, more generally,* $f^{p^k} = f(X^{p^k})$ *for all* $k \geq 0$.

Proof. (a) The equality $(f \cdot g)^p = f^p \cdot g^p$ is a direct consequence of commutativity of multiplication in $\mathbb{Z}_p[X]$. For addition, the binomial theorem for the ring $\mathbb{Z}_p[X]$ (see (A.1.7) in the appendix) says that

$$(f + g)^p = f^p + \sum_{1 \leq j \leq p-1} \binom{p}{j} f^j \cdot g^{p-j} + g^p. \tag{7.1.2}$$

All factors in the sum are to be reduced modulo the prime number p. Now for $1 \leq j \leq p - 1$ in the binomial coefficient

$$\binom{p}{j} = \frac{p(p-1)\cdots(p-j+1)}{j \cdot (j-1)\cdots 2 \cdot 1}$$

the numerator is divisible by p (since $j \geq 1$), but the denominator is not, since p is a prime number and $j \leq p - 1$. Thus, seen modulo p, the whole sum in (7.1.2) vanishes, and we have $(f + g)^p = f^p + g^p$.
(b) Let $f = a_d \cdot X^d + \cdots + a_1 \cdot X + a_0$. We apply (a) repeatedly to obtain

$$f^p = a_d^p \cdot (X^d)^p + \cdots + a_1^p \cdot X^p + a_0^p.$$

Now recall Fermat's Little Theorem (Theorem 4.2.10) to note that $a_i^p \equiv a_i$ (mod p) for $0 \leq i \leq d$, and exchange factors in the exponents to conclude that

$$f^p = a_d \cdot (X^p)^d + \cdots + a_1 \cdot X^p + a_0 = f(X^p).$$

For the more general exponent p^k we use induction. For $k = 0$, there is nothing to prove, since $f = f(X)$. The case $k = 1$ has just been treated. Now assume $k \geq 2$, and calculate in $\mathbb{Z}_p[X]$, by using the induction hypothesis and the case $k = 1$:

$$f^{p^k} = \left(f^{p^{k-1}}\right)^p = (f(X^{p^{k-1}}))^p = f((X^p)^{p^{k-1}}) = f(X^{p^k}),$$

as desired.   □

7.2 Division with Remainder and Divisibility for Polynomials

Let R be a ring. Even if R is a field, the ring $R[X]$ is never a field: for every $f = (a_0, a_1, \ldots)$ we have $X \cdot f = (0, a_0, a_1, \ldots) \neq (1, 0, 0, \ldots) = 1$, and hence X does not have a multiplicative inverse in $R[X]$. We will soon see how to use polynomials to construct fields. However, for many polynomials we may carry out a division with remainder, just as for integers.

Proposition 7.2.1 (Polynomial Division with Remainder). *Let R be a ring, and let $h \in R[X]$ be a monic polynomial (or a nonzero polynomial whose leading coefficient is a unit in R). Then for each $f \in R[X]$ there are unique polynomials $q, r \in R[X]$ with $f = h \cdot q + r$ and $\deg(r) < \deg(h)$.*

Proof. "Existence": First assume that h is monic, and write $h = (a_0, \ldots, a_d)$ with $a_d = 1$. We prove the existence of the quotient-remainder representation $f = h \cdot q + r$ by induction on $d' = \deg(f)$. If $d' < d$, then $q = 0$ and $r = f$ satisfy the claim. For the induction step we may assume that $d' \geq d \geq 0$ and that the claim is true for all polynomials f_1 with degree $d'_1 < d'$ in place of f. Write $f = (b_0, \ldots, b_{d'})$. Let

$$f_1 = f - b_{d'} \cdot X^{d'-d} \cdot h,$$

where $h_1 - h_2$ denotes the element $h_1 + (-h_2)$ in $R[X]$. Then

$$f_1 = (b'_0, \ldots, b'_{d'-1}, b_{d'} - b_{d'} \cdot 1),$$

for certain $b'_0, \ldots, b'_{d'-1}$, hence the degree of f_1 is smaller than d'. By the induction hypothesis, or algorithmically, by iterating the process, we see that f_1 can be written as $h \cdot q_1 + r$ for a polynomial r with $\deg(r) < d$. If we let $q = b_{d'} \cdot X^{d'-d} + q_1$, then $f = h \cdot q + r$, as desired.

If, more generally, $h = (a_0, \ldots, a_d)$ with a_d a unit, then let u be such that $u \cdot a_d = 1$, and consider the monic polynomial $h_1 = u \cdot h$ (see Lemma 7.1.9). By the above, we can write $f = h_1 \cdot q + r = h \cdot (u \cdot q) + r$ for some $q, r \in R[X]$ with $\deg(r) < \deg(h_1) = \deg(h)$.

"Uniqueness": *Case* 1: $f = 0$, i.e., $h \cdot q + r = 0$ with $\deg(r) < \deg(h)$.
— Obviously, then, $\deg(r) = \deg(h \cdot q)$. Since h has a unit as its leading
coefficient, we may apply Lemma 7.1.9(c) to conclude that $\deg(h) > \deg(r) = \deg(h \cdot q) = \deg(h) + \deg(q)$. This is only possible if $\deg(q) = -\infty$, i.e., $q = 0$,
and hence $r = 0$ as well. This means that $f = h \cdot 0 + 0$ is the only way to
split the zero polynomial in the required fashion. (The reader should make
sure he or she understands that if h has coefficients that are zero divisors, it
is well possible to write $0 = h \cdot q$ for a nonzero polynomial q.)
Case 2: $f \neq 0$, and $f = h \cdot q_1 + r_1 = h \cdot q_2 + r_2$. — Then $0 = h \cdot (q_1 - q_2) + (r_1 - r_2)$,
with $\deg(r_1 - r_2) \leq \max\{\deg(r_1), \deg(r_2)\} < \deg(h)$, and hence, by Case 1,
$q_1 - q_2 = r_1 - r_2 = 0$, which means that $q_1 = q_2$ and $r_1 = r_2$.   □

As an example, we consider a division with remainder in $\mathbb{Z}_{15}[X]$. That
is, all calculations in the following are carried out modulo 15. Consider the
polynomials $f = 4X^4 + 5X^2 + 6X + 1$ and $h = X^2 + 6$, which is monic.
Calculating modulo 15, we see that

$$f - 4X^2 \cdot h = 4X^4 + 5X^2 + 6X + 1 - (4X^4 + 9X^2) = 11X^2 + 6X + 1 = f_1.$$

Further,

$$f_1 - 11 \cdot h = 11X^2 + 6X + 1 - (11X^2 + 6) = 6X + 10 = f_2.$$

Putting these equations together, we can write

$$f = (4X^2 + 11) \cdot h + (6X + 10),$$

yielding the desired quotient-remainder representation.

Here is an iterative formulation of the algorithm for polynomial division
as indicated by the "existence" part of the proof of Proposition 7.2.1.

Algorithm 7.2.2 (Polynomial Division)

INPUT: Two polynomials over ring R:
        f (coefficients in f[0..d']) and
        h (coefficients in h[0..d] ; h[d] is a unit) .

METHOD:
```
1     q[0..d' − d], r[0..d − 1] : array of R ;
2     a : R;
3     find the unique u ∈ R with u · h[d] = 1;
4     for i from d' downto d do
5         a ← u · f[i];
6         q[i − d] ← a;
7         for j from i downto i − d do
8             f[j] ← f[j] − a · h[j] ;
9     for i from 0 to d − 1 do r[i] ← f[i];
10    return (q[0..d' − d]), (r[0..d − 1]);
```

In the execution of the j-loop in lines 7–8 in which the content of i is i, the polynomial $\texttt{f}[i] \cdot u \cdot h \cdot X^{i-d}$ is subtracted from $(\texttt{f}[0], \ldots, \texttt{f}[d'])$. This causes $\texttt{f}[i]$ to attain the value 0. In line 5, the same polynomial is added to $h \cdot (\texttt{q}[0], \ldots, \texttt{q}[d'-d])$. This means that one execution of the i-loop leaves the polynomial $(\texttt{f}[0], \ldots, \texttt{f}[d']) + h \cdot (\texttt{q}[0], \ldots, \texttt{q}[d'-d])$ unaltered. As this loop is carried out for $i = d', d'-1, \ldots, d$, after the execution of the i-loop is completed, $\texttt{f}[d], \ldots, \texttt{f}[d']$ all have become 0. The number of operations in R needed for this procedure is $O((d'-d)d)$.

Note that the method can be implemented in a more efficient way if h has only few nonzero coefficients, because then instead of the j-loop in lines 7 and 8 one will use a loop that only touches the nonzero entries of h. In the deterministic primality testing algorithm in Chap. 8 we will use $h = X^r - 1$ for some r; instead of the j-loop only one addition is needed.

Definition 7.2.3. *For $f, h \in R[X]$, we say that h **divides** f (or h is a **divisor** of f or f **is divisible by** h) if $f = h \cdot q$ for some $q \in R[X]$. If $0 < \deg(h) < \deg(f)$, then h is called a **proper divisor** of f.*

The reader should be warned that for arbitrary rings that contain zero divisors, this relation may have slightly surprising properties. For example, in $\mathbb{Z}_{12}[X]$ we have $(6X^2+4)(6X^2+2) = (6X^2+4)(6X^3+8) = 4$, so $6X^2+4$ divides 4 in this ring, and the "quotient" is not uniquely determined. However, by Lemma 7.1.9 and Proposition 7.2.1, if the leading coefficient of h is a unit such strange effects do not occur: if $f = h \cdot q$ then $\deg(f) = \deg(h) + \deg(q)$ and q is uniquely determined.

Definition 7.2.4. *Let $h \in R[X]$ be a polynomial whose leading coefficient is a unit. For $f, g \in R[X]$ we say that f and g are **congruent modulo** h, in symbols $f \equiv g \pmod{h}$, if $f - g$ is divisible by h.*

Note that it is sufficient to consider monic polynomials as divisors h in this definition, since h and $h_1 = u \cdot h$ create the same congruence relation, if u is an arbitrary unit in R.

Another way of describing the congruence relation is to say that $g = f + h \cdot q$, for some (uniquely determined!) polynomial q. It is very easy to check that the relation $f \equiv g \pmod{h}$ is an equivalence relation. (h divides $f - f = 0$; if h divides $f - g$ then h divides $g - f$; if h divides $f_1 - f_2$ and $f_2 - f_3$, then h divides $f_1 - f_3 = (f_1 - f_2) + (f_2 - f_3)$.) This means that this relation splits $R[X]$ into disjoint equivalence classes. Further, the relation fits together with the arithmetic operations in $R[X]$ and with the operation of substituting polynomials in polynomials:

Lemma 7.2.5. *Assume $h \in R[X] - \{0\}$ is a monic polynomial, and $f_1 \equiv f_2$ (mod h) and $g_1 \equiv g_2$ (mod h). Then*

(a) $f_1 + g_1 \equiv f_2 + g_2 \pmod{h}$;
(b) $f_1 \cdot g_1 \equiv f_2 \cdot g_2 \pmod{h}$;
(c) $f(g_1) \equiv f(g_2) \pmod{h}$ for all $f \in R[X]$.

Proof. Assume $f_1 - f_2 = h \cdot q_f$ and $g_1 - g_2 = h \cdot q_g$.

(a) and (b) We calculate: $(f_1 + g_1) - (f_2 + g_2) = h \cdot (q_f + q_g)$ and $(f_1 \cdot g_1) - (f_2 \cdot g_2) = (f_1 - f_2) \cdot g_1 + f_2 \cdot (g_1 - g_2) = h \cdot (q_f \cdot g_1 + f_2 \cdot q_g)$.

(c) If $f = a$ for some $a \in R$ or if $f = X$, there is nothing to show. By applying (b) repeatedly, we obtain the claim for all monomials $f = a \cdot X^s$, for $a \in R$ and $s \geq 0$. Using this, and applying (a) repeatedly, we get the claim for all polynomials $f = a_d X^d + \cdots + a_1 X + a_0$. $\qquad\square$

Lemma 7.2.5 means that in expressions to be transformed modulo h we may freely substitute subexpressions for one another as long as these are congruent modulo h.

We note that, just as within the integers, a divisor of h creates a coarser equivalence relation than congruence modulo h.

Lemma 7.2.6. *Assume $h, h' \in R[X]$ are monic polynomials, and assume that h' divides h. Then for all $f, g \in R[X]$ we have*

$$f \equiv g \pmod{h} \quad \Rightarrow \quad f \equiv g \pmod{h'}.$$

Proof. Write $h = \hat{h} \cdot h'$, and assume $f - g = q \cdot h$. Then $f - g = q \cdot (\hat{h} \cdot h') = (q \cdot \hat{h}) \cdot h'$, hence $f \equiv g \pmod{h'}$. $\qquad\square$

Now we are looking for canonical representatives of the equivalence classes induced by the congruence relation modulo h. We find them among the polynomials whose degree is smaller than the degree of h.

Lemma 7.2.7. *Assume $h \in R[X]$, $d = \deg(h) \geq 0$, is a monic polynomial. Then for each $f \in R[X]$ there is exactly one $r \in R[X]$, $\deg(r) < d$, with $f \equiv r \pmod{h}$.*

Proof. From Proposition 7.2.1 we get that for given f there are uniquely determined polynomials q and r with $\deg(r) < d$ such that $f = h \cdot q + r$, or $f - r = h \cdot q$. This statement is even stronger than the claim of the lemma. $\quad\square$

The remainder polynomial r is called $f \bmod h$, in analogy to the notation for the integers. We note that from Lemma 7.2.7 it is immediate that $f \bmod h = g \bmod h$ holds if and only if $f \equiv g \pmod{h}$. Further, from Lemma 7.2.5 it is immediate that $(f + g) \bmod h = ((f \bmod h) + (g \bmod h)) \bmod h$ and $(f \cdot g) \bmod h = ((f \bmod h) \cdot (g \bmod h)) \bmod h$.

7.3 Quotients of Rings of Polynomials

Now we have reached a position in which we can define another class of structures, which will turn out to be central in the deterministic primality test. Just as in the case of integers taken modulo some m, we can take the elements of $R[X]$ modulo some polynomial h, and define a ring structure on the set of remainder polynomials.

Definition 7.3.1. *If $(R, +, \cdot, 0, 1)$ is a ring, and $h \in R[X]$, $d = \deg(h) \geq 0$, is a monic polynomial, we let $R[X]/(h)$ be the set of all polynomials in $R[X]$ of degree strictly smaller than d, together with the following operations $+_h$ and $\cdot_h$:*

$$f +_h g = (f + g) \bmod h \text{ and } f \cdot_h g = (f \cdot g) \bmod h, \text{ for } f, g \in R[X]/(h).$$

Example 7.3.2. Let us consider the polynomial $h = X^4 + 3X^3 + 1$ from $\mathbb{Z}_{12}[X]$. The ring $\mathbb{Z}_{12}[X]/(h)$ consists of the $12^4 = 20736$ polynomials over $\mathbb{Z}_{12}$ of degree up to 3. To multiply $f = 2X^3$ and $g = X^2 + 5$ we calculate the product $f \cdot g = 2X^5 + 10X^3$ and determine the remainder modulo h as indicated in the proof of Proposition 7.2.1. (Calculations in $\mathbb{Z}_{12}$ are carried out without comment.)

$$2X^5 + 10X^3 \equiv 2X^5 + 10X^3 - 2X \cdot h \equiv 6X^4 + 10X^3 + 10X$$
$$\equiv 6X^4 + 10X^3 + 10X - 6 \cdot h \equiv 4X^3 + 10X + 6 \pmod{h}.$$

Thus, $2X^3 \cdot_h (X^2 + 5) = 4X^3 + 10X + 6$. Using coefficient vectors of length 4, this would read $(0, 0, 0, 2) \cdot_h (5, 0, 1, 0) = (6, 10, 0, 4)$.

Proposition 7.3.3. *If R and h are as in the preceding definition, then $(R[X]/(h), +_h, \cdot_h, 0, 1)$ is a ring with 1. Moreover, we have:*

(a) *$f \bmod h = f$ if and only if $\deg(f) < d$;*
(b) *$(f + g) \bmod h = ((f \bmod h) + (g \bmod h)) \bmod h$ and $(f \cdot g) \bmod h = ((f \bmod h) \cdot (g \bmod h)) \bmod h$, for all $f, g \in R[X]$;*
(c) *If $g_1 \equiv g_2 \pmod{h}$, then $f(g_1) \bmod h = f(g_2) \bmod h$, for all $f, g_1, g_2 \in R[X]$.*

Proof. We must check the conditions in Definition 4.3.2.
By definition, $R[X]/(h)$ is closed under addition $+_h$ and multiplication $\cdot_h$. Clearly the zero polynomial acts as a neutral element for $+_h$, and the polynomial 1 acts as a neutral element for $\cdot_h$. Finally, the polynomial $-f \in R[X]/(h)$ is an inverse for $f \in R[X]/(h)$ with respect to $+_h$. To check the distributive law is a matter of routine, using the remarks after Lemma 7.2.7. Claim (a) is obvious. Claims (b) and (c) follow from Lemmas 7.2.7 and 7.2.5.   □

The reader should note that the ground set $R[X]/(h)$ does not depend on h, but only on $\deg(h)$. It is the operations $+_h$ and $\cdot_h$ that truly depend on h.

Remark 7.3.4. It is no problem to implement the structure $R[X]/(h)$ and its operations on a computer. We just indicate the principles of such an implementation. The elements of $R[X]/(h)$ are represented as arrays of length d. Adding two such elements can trivially be done by performing d additions in R. Multiplying two polynomials f and g can be done in the naive way by performing d^2 multiplications and $(d-1)^2$ additions in R, or by using faster methods with $O(d(\log d)(\log \log d))$ ring operations; see Remark 7.1.8. Finally, we calculate $(f \cdot g) \bmod h$ by the procedure for polynomial division from the previous section (we may even omit the operations that build up the quotient polynomial). It is obvious that overall $O(d^2)$ multiplications and additions in R are performed.

For very small structures $R[X]/(h)$, the arithmetic rules can be listed in tables, so that one obtains an immediate impression as to what these operations look like. As an example, let us consider the tables that describe arithmetic in $\mathbb{Z}_3[X]/(h)$ for $h = X^2 + 1$. That is, the integers are taken modulo 3, and the underlying set is just the set of polynomials of degree ≤ 1, that is, linear and constant polynomials. For compactness, we represent a polynomial $a + bX$ by its coefficient sequence ab. (Since the coefficients are only $0, 1$, and 2, no parentheses are needed.) The zero polynomial is 00, the polynomial X is 01, and so on. There are nine polynomials in the structure. To write out the addition table in full would yield a boring result, since this is just addition modulo 3 in both components, e.g., $22 +_h 21 = 10$.

$\cdot_h$	00	10	20	01	11	21	02	12	22
00	00	00	00	00	00	00	00	00	00
10	00	10	20	01	11	21	02	12	22
20	00	20	10	02	22	12	01	21	11
01	00	01	02	20	21	22	10	11	12
11	00	11	22	21	02	10	12	20	01
21	00	21	12	22	10	01	11	02	20
02	00	02	01	10	12	11	20	22	21
12	00	12	21	11	20	02	22	01	10
22	00	22	11	12	01	20	21	10	02

Table 7.1. Multiplication table of $\mathbb{Z}_3[X]/(X^2 + 1)$

For multiplication, we recall that modulo 3 we have $0 \cdot x = 0$, $1 \cdot x = x$, and $2 \cdot 2 = 1$. Now we determine the entries in the multiplication table row by row (the result can be found in Table 7.1). The multiples of the zero polynomial 00 are all 00, and $10 \cdot_h f = f$ for all f. Further, $20 \cdot_h f$ results from f by doubling all coefficients. The first interesting case is $01 \cdot_h 01 = X \cdot_h X = X^2 \bmod h$. Since $X^2 \equiv 2 \pmod{X^2 + 1}$, we get $01 \cdot_h 01 = 20$, and hence $01 \cdot_h 02 = 10$. Next, $01 \cdot_h 11 = X^2 + X \pmod{h}$. Again replacing X^2 by 2, we obtain $01 \cdot_h 11 = 21$, and hence, by doubling, $01 \cdot_h 22 = 12$. Further, $01 \cdot_h 21 = 01 \cdot_h 20 +_h 01 \cdot_h 01 = 02 +_h 20 = 22$. Continuing in this way, we complete the row for 01. The row for 11 is obtained by adding corresponding entries in the row for 10 and that for 01, similarly for the row for 21. To obtain the row for 02, we may double (modulo 3) the entries in the row for 01, and obtain the rows for 12 and 22 again by addition. (Clearly, if a different polynomial had been used, we would obtain a table of the same kind, with different entries.)

This little example should make it clear that although the definition of the structure $R[X]/(h)$ may look a little obscure and complicated at first glance, it really yields a very clean structure based on the simple set of all d-tuples from R.

7.4 Irreducible Polynomials and Factorization

In this section, we consider polynomials over a field F. The standard examples, which the reader should have in mind, are the fields $\mathbb{Q}$ and $\mathbb{Z}_p$, for p an arbitrary prime number.

Definition 7.4.1. *A polynomial $f \in F[X] - \{0\}$ is called **irreducible** if f does not have a proper divisor, i.e., if from $f = g \cdot h$ for $g, h \in F[X]$ it follows that $g \in F^*$ or $\deg(g) = \deg(f)$.*

This means that the only way to write f as a product is trivial: as $f = (a_0, \ldots, a_d) = a \cdot (b_0, \ldots, b_d)$, where b_i results from a_i by multiplication with the field element a^{-1}.

As an example, consider some polynomials in $\mathbb{Z}_5[X]$: $f_1 = X^2 + 4X + 1$ is irreducible, since it is impossible to write $f_1 = g \cdot h$ with $\deg(g) = \deg(h) = 1$. (Assume we could write $f_1 = (a + bX) \cdot (a' + b'X)$, with $b \neq 0$. Then the field element $c = -a \cdot b^{-1}$ would satisfy $f_1(c) = 0$. But we can check directly that no element $c \in \mathbb{Z}_5$ satisfies $f_1(c) = 0$.) On the other hand, $f_1 = 2 \cdot (3X^2 + 2X + 3)$ is a way of writing f_1 as a (trivial) product. — Similarly, $f_2 = 2X^3 + X + 4$ is irreducible: there is no way of writing it as a product $(a + bX) \cdot (a' + b'X + c'X^2)$ with $b \neq 0$, again since f_2 does not have a root in $\mathbb{Z}_5$. — Next, consider $f = 3X^3 + 2X^2 + 4X + 1$. This polynomial is not irreducible, since $f = (X + 4) \cdot (3X^2 + 4) = 3 \cdot (X + 4) \cdot (X^2 + 4)$.

It should be clear that the notion of irreducibility depends on the underlying field. For example, for $F = \mathbb{Q}$, the polynomial $X^2 + 1$ is irreducible (since it has no roots in $\mathbb{Q}$), while for $F = \mathbb{Z}_2$ we have $X^2 + 1 = (X + 1)(X + 1)$ (the coefficients are elements of $\mathbb{Z}_2$).

We call two polynomials f and g from $F[X]$ **associated** if $f = a \cdot g$ for some $a \in F^*$. It is obvious that this defines an equivalence relation on the set of nonzero polynomials, and that in each equivalence class there is exactly one monic polynomial, which we regard as the canonical representative of the class. Clearly, either all or none of the elements of an equivalence class are irreducible.

Theorem 7.4.4, to follow below, essentially says that in $F[X]$ the irreducible polynomials play the same role as the prime numbers among the integers: polynomials over a field F have the *unique factorization property*, i.e., every polynomial can be written as a product of irreducible polynomials in essentially one way. To prepare for the proof, we need two lemmas. The reader will notice that these lemmas correspond to statements that are true in the ring $\mathbb{Z}$, if one replaces "irreducible polynomial" with "prime number"; see Propositions 3.1.13 and 3.1.15.

Lemma 7.4.2. *Let $h \in F[X]$ be irreducible, and let $f \in F[X]$ be such that h does not divide f. Then there are polynomials s and t such that*

$$1 = s \cdot h + t \cdot f.$$

Proof. Let $I = \{s \cdot h + t \cdot f \mid s, t \in F[X]\}$. Clearly, $0 \in I$, and $I - \{0\} \neq \emptyset$. Let $g = s \cdot h + t \cdot f$ be an element of $I - \{0\}$ of minimal degree, and let $d = \deg(g) \geq 0$.

Claim: g divides f and g divides h.

Proof of Claim: By Proposition 7.2.1, $f = g \cdot q + r$ for uniquely determined polynomials q and r, where $\deg(r) < \deg(g)$. Since $r = f - g \cdot q = (-s \cdot q) \cdot h + (1 - t \cdot q) \cdot f \in I$, the minimality of $\deg(g)$ implies that $r = 0$, or $f = g \cdot q$. — That g divides h is proved in exactly the same way.

Because of the claim we may write $h = g \cdot q$ for some polynomial q. But h is irreducible, so there are only two possibilities:

Case 1: $g = a$ for some $a \in F^*$. — Then $1 = a^{-1} \cdot g = (a^{-1} \cdot s) \cdot h + (a^{-1} \cdot t) \cdot f$, as desired.

Case 2: $q = b$ for some $b \in F^*$. — By the claim, we may write $f = g \cdot q'$ for some q'. Then $f = q' \cdot g = q' \cdot (b^{-1} \cdot h) = (b^{-1} \cdot q') \cdot h$, contradicting the assumption that h does not divide f. Hence this case cannot occur.   □

Lemma 7.4.3. *Let $h \in F[X]$ be irreducible. If $f \in F[X]$ is divisible by h and $f = g_1 \cdot g_2$, then h divides g_1 or h divides g_2.*

Proof. Write $f = h \cdot q$. Assume that h does not divide g_1. By the preceding lemma, there are polynomials s and t such that $1 = s \cdot h + t \cdot g_1$. If we multiply this equation by g_2, we obtain

$$g_2 = g_2 \cdot s \cdot h + t \cdot f = g_2 \cdot s \cdot h + t \cdot h \cdot q,$$

so g_2 is divisible by h.   □

Theorem 7.4.4 (Unique Factorization for Polynomials). *Let F be a field. Then every nonzero polynomial $f \in F[X]$ can be written as a product $a \cdot h_1 \cdots h_s$, $s \geq 0$, where $a \in F^*$ and $h_1, \ldots, h_s$ are monic irreducible polynomials in $F[X]$ of degree > 0. This product representation is unique up to the order of the factors.*

Proof. "Existence": We use induction on the degree of f. If $f = a \in F - \{0\}$, then $f = a \cdot 1$ for the empty product 1. Now assume that $d = \deg(f) > 0$, and that the claim is true for all polynomials of degree $< d$. Write $f = a_d X^d + \cdots + a_1 X + a_0$. If f is irreducible, we let $a = a_d$ and $h_1 = a^{-1} \cdot f$. Now assume that $f = g_1 \cdot g_2$ for polynomials g_1 and g_2 with $\deg(g_1), \deg(g_2) > 0$. By Lemma 7.1.9(c) we have that $\deg(f) = \deg(g_1) + \deg(g_2)$, hence

$\deg(g_1), \deg(g_2) < d$. By the induction hypothesis, we may write g_1 and g_2 as products of a field element and monic irreducible polynomials. Putting these two products together yields the desired representation for f.

"Uniqueness": This is proved indirectly. Assume for a contradiction that there are nonzero polynomials with two essentially distinct factorizations. Let f be one such polynomial with minimal degree. Assume

$$f = a \cdot h_1 \cdots h_s = a' \cdot h'_1 \cdots h'_t$$

are two different factorizations. Clearly, then, $s, t \geq 1$. Now h'_1 is a divisor of $a \cdot (a')^{-1} \cdot h_1 \cdots h_s$. By applying Lemma 7.4.3 repeatedly, we see that h'_1 must divide h_j for some j, $1 \leq j \leq s$. By reordering the h_j, we can arrange it so that $j = 1$. Thus, write $h_1 = h'_1 \cdot q$. Because h_1 is irreducible and h'_1 is not in F^*, we must have $q \in F^*$. From the assumption that h'_1 and h_1 are monic, we conclude $h'_1 = h_1$. This means that

$$a \cdot h_2 \cdots h_s = a' \cdot h'_2 \cdots h'_t$$

are two different factorizations of a polynomial of degree smaller than $\deg(f)$, contradicting our choice of f. — This shows that there are no polynomials with two different factorizations.   □

It should be mentioned that in general no polynomial time algorithm is known that can find the representation of a polynomial f as a product of irreducible factors. Special cases (like $F[X]$ where F is a field with a small cardinality) can be treated quite efficiently. However, for our purposes, no such algorithm is needed, since the factorization of polynomials is not used algorithmically, but only as a tool in the correctness proof of the deterministic primality test.

The concept of an irreducible polynomial is central in a method for constructing finite fields other than the fields $\mathbb{Z}_p$.

Theorem 7.4.5. *Let F be a field, and let $h \in F[X]$ be a monic irreducible polynomial over F. Then the structure $F[X]/(h)$ from Definition 7.3.1 is a field. (If F is finite, this field has $|F|^{\deg(h)}$ elements.)*

Proof. We have seen already in Proposition 7.3.3 that $F[X]/(h)$ is a ring. It remains to show that every element of $F[X]/(h) - \{0\}$ has an inverse with respect to multiplication modulo h. But this we have proved already: Let $f \in F[X]/(h) - \{0\}$ be arbitrary. Since $\deg(f) < \deg(h)$, it is not possible that h divides f. By Lemma 7.4.2 there are polynomials $s, t \in F[X]$ so that $1 = s \cdot h + t \cdot f$. This means that $1 \equiv t \cdot f \equiv (t \bmod h) \cdot f \pmod{h}$, i.e., $t \bmod h$ is a multiplicative inverse of f in $F[X]/(h) - \{0\}$.   □

As an example, the reader may go back and have another look at the multiplication table of the ring $\mathbb{Z}_3[X]/(X^2 + 1)$ (Table 7.1). Each of the eight nonzero elements has a multiplicative inverse, as is immediately read off from the table. For example, the entry 01, which corresponds to the polynomial

X, has the inverse 02, which corresponds to $2X$. Indeed, $X \cdot 2X = 2X^2 = 2 \cdot (-1) = 2 \cdot 2 \equiv 1$, calculated in $\mathbb{Z}_3[X]/(X^2+1)$. Similarly, in $\mathbb{Z}_5[X]/(X^2 + 4X + 1)$, the polynomial X has the inverse $(4X + 1)$, since $X \cdot (4X + 1) = 4X^2 + X \equiv 4(X^2 + 4X + 1) + 1 \equiv 1$.

It should be mentioned that (in contrast to the more difficult task of factoring polynomials into irreducible factors) a version of the Extended Euclidean Algorithm 3.2.4 known from the integers yields an efficient procedure to calculate inverses in the field $F[X]/(h) - \{0\}$. Since this algorithm is not relevant for our task, we do not describe it here.

The question remains if there are sufficiently many irreducible polynomials to make Theorem 7.4.5 a useful approach to obtaining finite fields. It is a well-known fact, to be proved by methods not described in this book, that for every field F and for every $d \geq 0$ there is at least one irreducible (monic) polynomial of degree d over F, which then leads to the construction of a field that consists of d-tuples of elements of F. If F is finite, the cardinality of this field is $|F|^d$. Starting with the fields $\mathbb{Z}_p$ for p a prime number, we obtain fields of cardinality p^d for every prime number p and every exponent d. In the other direction, it is not hard to show by basic methods from linear algebra that if there is a finite field of cardinality q then q is the power of a prime number p.

The field $F[X]/(h)$ has the interesting property that it contains a root of the polynomial h. This fact will be very important later.

Proposition 7.4.6. *Let F and h be as in the previous theorem, and let $F' = F[X]/(h)$ be the corresponding field. Then the element $\zeta = X \bmod h \in F'$ is a root of h, i.e., in F' we have $h(\zeta) = 0$.*
(Note that if $\deg(h) \geq 2$ then $\zeta = X \in F' - F$. If $\deg(h) = 1$, then $h = X + a$ for some $a \in F'$ and $\zeta = -a$.)

Proof. We use Proposition 7.3.3 for calculating modulo h in $F[X]$, and Example 7.1.14(b) to obtain

$$h(\zeta) = h(X \bmod h) \bmod h = h(X) \bmod h = h \bmod h = 0. \qquad \square$$

7.5 Roots of Polynomials

From calculus it is well known that if we consider nonzero polynomials over $\mathbb{R}$, then linear functions $x \mapsto ax+b$ have at most one root, quadratic functions $x \mapsto ax^2 + bx + c$ have at most two, cubic polynomials have at most three, and so on. We note here that this is a property that holds in all fields. The basis for this observation is simply division with remainder, which shows that if a is a root of f then f contains $X - a$ as a factor.

Theorem 7.5.1. *Let F be a field, and let $f \in F[X]$ with $f \neq 0$, i.e., $d = \deg(f) \geq 0$. Then*

$$|\{a \in F \mid f(a) = 0\}| \leq d.$$

Proof. We proceed by induction on d. If $d = 0$, f is an element b of $F - \{0\}$, which by Proposition 7.1.13(a) means that $f(a) = b$ for all a, so that f has no root at all. For the induction step, assume $d \geq 1$. If f has no root, we are done. If $f(a) = 0$ for some $a \in F$, we use Proposition 7.2.1 to write

$$f = (X - a) \cdot f_1 + r,$$

where $\deg(r) < \deg(X - a) = 1$, hence $r \in F$. By substituting a in both sides, see Proposition 7.1.13(c), we obtain

$$0 = f(a) = (a - a) \cdot f_1(a) + r = r,$$

hence in fact $f = (X - a) \cdot f_1$. By Lemma 7.1.9(c), $\deg(f_1) = d - 1$. Applying the induction hypothesis to f_1 we get that the set $A_1 = \{a \in F \mid f_1(a) = 0\}$ has at most $d-1$ elements. It remains to show that all roots of f are contained in $A = A_1 \cup \{a\}$. But this is clear, again using Proposition 7.1.13(c): If $f(b) = (b - a) \cdot f_1(b) = 0$, then $b - a = 0$ or $f_1(b) = 0$.   □

Corollary 7.5.2. *If* $\deg(f), \deg(g) \leq d$ *and there are* $d + 1$ *elements* $b \in F$ *with* $f(b) = g(b)$, *then* $f = g$.

Proof. Consider $h = f - g$. From the assumption it follows that $\deg(h) \leq d$ and that $h(b) = 0$ for $d + 1$ elements of F. From Theorem 7.5.1 we conclude that $h = 0$, hence $f = g$.   □

7.6 Roots of the Polynomial $X^r - 1$

The polynomial $X^r - 1$ over a ring $\mathbb{Z}_m[X]$, together with its irreducible factors, is at the center of interest in the correctness proof of the deterministic primality test in Chap. 8. For later use, we state some of its properties.

Observe first that

$$X^r - 1 = (X - 1)(X^{r-1} + \cdots + X + 1), \tag{7.6.3}$$

for $r \geq 1$. This equation holds over any ring with 1, as can be checked by multiplying out the right-hand side. Further, the following simple generalization will be helpful:

$$X^{rs} - 1 = (X^r - 1)((X^r)^{s-1} + (X^r)^{s-2} + \cdots + X^r + 1), \tag{7.6.4}$$

i.e., $X^r - 1$ divides $X^{rs} - 1$, for $r, s \geq 1$. In particular, these equalities are true in $\mathbb{Z}_m[X]$ for arbitrary $m \geq 2$. Whether and how the second factor $X^{r-1} + \cdots + X + 1$ in (7.6.3) splits into factors depends on r and on m. We focus on the case where r and $m = p$ are different prime numbers. In this case, we know from Theorem 7.4.4 that in $\mathbb{Z}_p[X]$ the polynomial $X^{r-1} + \cdots + X + 1$ splits into monic irreducible factors that are uniquely determined.

Example 7.6.1. (a) In $\mathbb{Z}_{11}[X]$ we have for $r = 5$ that $X^4 + \cdots + X + 1 = (X + 8)(X + 7)(X + 6)(X + 2)$; the polynomial $\frac{X^r - 1}{X - 1}$ splits into linear factors.

(b) In $\mathbb{Z}_{11}[X]$ we have for $r = 7$ that $X^6 + \cdots + X + 1 = (X^3 + 5X^2 + 4X + 10)(X^3 + 7X^2 + 6X + 10)$; the two factors are irreducible.

(c) In $\mathbb{Z}_7[X]$ we have for $r = 5$ that $X^4 + \cdots + X + 1$ is irreducible.

We will see shortly in which way r and p determine what these irreducible factors of $X^{r-1} + \cdots + X + 1$ look like. In any case, we may take any one of these irreducible factors, h, say, and construct the field $\mathbb{Z}_p[X]/(h)$. This field has the crucial property that it contains a ***primitive rth root of unity***:

Proposition 7.6.2. *Let p and r be prime numbers with $p \neq r$, and let h be a monic irreducible factor of $\frac{X^r - 1}{X - 1} = X^{r-1} + \cdots + X + 1$. Then in the field $F = \mathbb{Z}_p[X]/(h)$ the element $\zeta = X \bmod h$ satisfies $\mathrm{ord}_F(\zeta) = r$.*

Proof. We may write $X^r - 1 = (X - 1) \cdot h \cdot q$, for some polynomial q. From Proposition 7.4.6 we know that $\zeta = X \bmod h$ is a root of h in F. Since h is a factor of $h \cdot q = X^{r-1} + \cdots + X + 1$ and of $X^r - 1$, the element ζ also is a root of these polynomials, and we get

$$\zeta^{r-1} + \cdots + \zeta + 1 = 0, \tag{7.6.5}$$

and

$$\zeta^r = 1, \tag{7.6.6}$$

in F. The last equation implies that the order of ζ in F is a divisor of r (see Proposition 4.2.7(b)). Now r is a prime number, so $\mathrm{ord}_F(\zeta) \in \{1, r\}$. Can it be 1? No, since this would mean that $\zeta = 1$, which would entail, by (7.6.5), that $1^{r-1} + 1^{r-2} + \cdots + 1^1 + 1 = 0$ in F, hence in $\mathbb{Z}_p$, i.e., $r \equiv 0 \pmod{p}$. This is impossible, since p is a prime number different from r. So the order of ζ is r. □

Remark 7.6.3. (a) The previous proposition implies that $1, \zeta, \ldots, \zeta^{r-1}$ are distinct. Clearly, for each j we have $(\zeta^j)^r = (\zeta^r)^j = 1$. This implies, by Theorem 7.5.1, that these r powers of ζ are exactly the r distinct roots of the polynomial $X^r - 1$ in F.

(b) In Proposition 7.4.6 we showed that $\zeta = X \bmod h$ is X if $\deg(h) > 1$ and is $-a \in \mathbb{Z}_p$ if $h = X + a$ has degree 1.

In the following proposition we determine exactly in which way $\frac{X^r - 1}{X - 1}$ splits into irreducible factors in $\mathbb{Z}_p[X]$. The crucial parameter is the order $\mathrm{ord}_r(p)$ of p in $\mathbb{Z}_r^*$. In Example 7.6.1 we have (a) $\mathrm{ord}_5(11) = 1$, (b) $\mathrm{ord}_7(11) = \mathrm{ord}_7(4) = 3$, (c) $\mathrm{ord}_5(7) = \mathrm{ord}_5(2) = 4$, as is easily checked.

Proposition 7.6.4. Let p and r be prime numbers with $p \neq r$, and $q = X^{r-1} + \cdots + X + 1$. Then

$$q = h_1 \cdots h_s,$$

where $h_1, \ldots, h_s \in \mathbb{Z}_p[X]$ are monic irreducible polynomials of degree $\mathrm{ord}_r(p)$.

Proof. By Theorem 7.4.4, we know that the factorization exists and that it is unique up to changing the order. Thus, it is sufficient to show the following:

(∗)   If $h \in \mathbb{Z}_p[X]$ is monic and irreducible, and divides $q = X^{r-1} + \cdots + X + 1$, then $\deg(h) = \mathrm{ord}_r(p)$.

Let $k = \mathrm{ord}_r(p)$. Clearly, $k \geq 1$. (Note that it could be possible that $p \equiv 1$ (mod r), i.e., that $k = 1$.) Further, let $h \in \mathbb{Z}_p[X]$ be a monic irreducible factor of q, of degree $d \geq 1$. We show that $k = d$. Again, we consider the field $F = \mathbb{Z}_p[X]/(h)$ of cardinality p^d.

"$\leq$": By Proposition 7.6.2, the multiplicative group of F contains an element of order r. Since the group has order $|F^*| = p^d - 1$, this implies (by Proposition 4.1.9) that r divides $p^d - 1$. In other words, $p^d \bmod r = 1$. Applying Proposition 4.2.7(b) to the multiplicative group $\mathbb{Z}_r^*$ yields that $k = \mathrm{ord}_r(p)$ is a divisor of d; this implies that $k \leq d$.

"$\geq$": We need an auxiliary statement about the field F.

Claim: $f^{p^k} = f$ for all $f \in F$.

Proof of Claim: Let $f \in F$ be arbitrary. That is, $f \in \mathbb{Z}_p[X]$ and $\deg(f) < d$. By Proposition 7.1.15 we have

$$f^{p^k} = f(X^{p^k}),\tag{7.6.7}$$

in $\mathbb{Z}_p[X]$. Since h is a divisor of $X^r - 1$ in $\mathbb{Z}_p[X]$, we have $X^r - 1 \equiv 0 \bmod h$, or $X^r \equiv 1 \bmod h$. The definition of $k = \mathrm{ord}_r(p)$ entails that $p^k \equiv 1$ (mod r), i.e., $p^k = mr + 1$ for some number m. Hence

$$X^{p^k} \equiv X^{mr+1} \equiv (X^r)^m \cdot X \equiv X \pmod{h}.$$

According to Proposition 7.3.3(c) we may substitute this into (7.6.7) and continue calculating modulo h to obtain

$$f^{p^k} \bmod h = f(X^{p^k}) \bmod h = f(X) \bmod h = f \bmod h = f.$$

But $f^{p^k} \bmod h$ is just f^{p^k} calculated in F, so the claim is proved.

By Theorem 4.4.3 the field F has a primitive element g, i.e., an element g of order $|F^*| = p^d - 1$. Applying the claim to g we see that in F we have $g^{p^k-1} = 1 = g^0$. Applying Proposition 4.2.7(b) we conclude that $p^k - 1$ is divisible by $p^d - 1$. This implies $k \geq d$, as desired.   □

8. Deterministic Primality Testing in Polynomial Time

In this chapter, we finally get to the main theme of this book: the deterministic polynomial time primality test of M. Agrawal, N. Kayal, and N. Saxena. The basis of the presentation given here is the revised version [4] of their paper "PRIMES is in P", as well as the correctness proof in the formulation of D.G. Bernstein [10].

This chapter is organized as follows. We first describe a simple characterization of prime numbers in terms of certain polynomial powers, which leads to the basic idea of the algorithm. Then the algorithm is given, slightly varying the original formulation. The time analysis is not difficult, given the preparations in Sect. 3.6. Some variations of the time analysis are discussed, one involving a deep theorem from analytic number theory, the other a number-theoretical conjecture. The main part of the chapter is devoted to the correctness proof, which is organized around a main theorem, as suggested in [10].

8.1 The Basic Idea

We start by explaining the very simple basic idea of the new deterministic primality testing algorithm. Consider the following characterization of prime numbers by polynomial exponentiation.

Lemma 8.1.1. *Let $n \geq 2$ be arbitrary, and let $a < n$ be an integer that is relatively prime to n. Then*

$$n \text{ is a prime number} \quad \Leftrightarrow \quad \text{in } \mathbb{Z}_n[X] \text{ we have } (X + a)^n = X^n + a.$$

Proof. We calculate in $\mathbb{Z}_n[X]$. By the binomial theorem (A.1.7) we have

$$(X + a)^n = X^n + \sum_{0 < i < n} \binom{n}{i} a^i X^{n-i} + a^n. \tag{8.1.1}$$

"$\Rightarrow$": (Cf. Proposition 7.1.15(a).) Assume that n is a prime number. Then for $1 \leq i \leq n - 1$ in the binomial coefficient

$$\binom{n}{i} = \frac{n(n-1)\cdots(n-i+1)}{i!}$$

M. Dietzfelbinger: Primality Testing in Polynomial Time, LNCS 3000, pp. 115-131, 2004.
© Springer-Verlag Berlin Heidelberg 2004

the numerator is divisible by n, but not the denominator, hence n divides $\binom{n}{i}$. This means that in $\mathbb{Z}_n[X]$ all these coefficients vanish. Further, in $\mathbb{Z}_n$ we have that $a^n = a$, by Fermat's Little Theorem (Theorem 4.2.10). Hence $(X + a)^n = X^n + a$, as desired.

"$\Leftarrow$": Assume that n is not a prime number, and choose a prime factor $p < n$ of n and some $s \geq 1$ such that p^s divides n but p^{s+1} does not. Consider the coefficient of X^{n-p} in (8.1.1):

$$\binom{n}{p} \cdot a^p = \frac{n(n - 1) \cdots (n - p + 1)}{p!} \cdot a^p.$$

The first factor n in the numerator of this fraction is divisible by p^s, the other factors are relatively prime to p. Hence the numerator is divisible by p^s, but not by p^{s+1}. The denominator is divisible by p. Because a and n are relatively prime, p does not divide a^p. Hence $\frac{n(n-1)\cdots(n-p+1)}{p!} \cdot a^p$ is not divisible by p^s, and hence not divisible by n. This means that $\binom{n}{p}a^p \not\equiv 0 \pmod{n}$, and hence $(X + a)^n \neq X^n + a$ in $\mathbb{Z}_n[X]$. $\qquad\square$

Lemma 8.1.1 suggests a simple method to test whether an odd number n is prime: Choose some $a < n$ that is relatively prime to n (e.g., $a = 1$). Calculate, by fast exponentiation in the ring $\mathbb{Z}_n[X]$, as described in Algorithm 4.3.9, the coefficients of the polynomial $(X + a)^n$ in $\mathbb{Z}_n[X]$. If the result is $X^n + a$, then n is a prime number, otherwise it is not.

The disadvantage of this test is that the polynomials produced in the course of the exponentiation procedure might have many nonzero terms. The degree of the polynomial to be squared in the last round of this procedure must be at least $(n - 1)/2$, and hence might have as many as $(n + 1)/2$ terms. Even if very efficient methods for polynomial multiplication are used, the number of arithmetic operations needed cannot be bounded by anything better than $O(n)$. This bound is much worse than if we used trial division by all odd numbers below $\sqrt{n}$.

To at least have the chance of obtaining an efficient algorithm, one tests the congruence

$$(X + a)^n \equiv X^n + a \tag{8.1.2}$$

not "absolutely" in $\mathbb{Z}_n[X]$, but modulo a polynomial $X^r - 1$, where r will have to be chosen in a clever way. That is, one compares, in $\mathbb{Z}_n[X]$, the polynomials

$$(X+a)^n \bmod (X^r - 1) \quad \text{and} \quad (X^n+a) \bmod (X^r - 1) = X^{n \bmod r}+a. \tag{8.1.3}$$

In the intermediate results that appear in the course of the computation of the power $(X + a)^n$, all coefficients are reduced modulo n, hence they can never exceed n. Calculating modulo $X^r - 1$ just means that one can replace X^s by $X^{s \bmod r}$, hence that the degrees of the polynomials that appear as intermediate results can be kept below r. This keeps the computational cost in the polynomial range as long as r is $O((\log n)^c)$ for some constant c.

If n is a prime number, the result $(X + a)^n \bmod (X^r - 1)$ will be equal to $X^{n \bmod r} + a$ for every a and for every r.

For the reverse direction it would be nice if it were sufficient to carry out the calculation of $(X + a)^n \bmod (X^r - 1)$ only for a single r, which should be not too large. (In the terminology of prior sections, such an r could be called an "*AKS-witness*" for the fact that n is a prime.) The crux of the deterministic primality test is to give a sufficient condition for numbers r to be suitable, which is not too hard to check, and to show that there are suitable r's that are of size $O((\log n)^c)$. Then finding such an r is a minor problem, since exhaustive search can be used. It turns out that even for suitable r it is not sufficient anymore to compare the two terms in (8.1.3) for one a, even if this a were cleverly chosen, but that these terms must be equal for a whole series of a's in order to allow one to conclude that n is ... no, not prime, but at least a *power* of a prime number. The case that n is a perfect power of any number can easily be excluded by a direct test.

In the course of searching for a witness r for n being prime the algorithm keeps performing tests on candidates for r that may prove immediately that n is composite: if an r is encountered so that r divides n (and $r < n$) or some r and some a are found so that the polynomials in (8.1.3) are different, the algorithm immediately returns the answer that n is composite, and r resp. the pair (r, a) is a witness to this fact.

8.2 The Algorithm of Agrawal, Kayal, and Saxena

Algorithm 8.2.1 (Deterministic Primality Test)

INPUT: Integer $n \geq 2$.

METHOD:

```
1      if ( n = aᵇ for some a, b ≥ 2 ) then return  "composite";
2      r ← 2;
3      while ( r < n ) do
4          if ( r divides n ) then return  "composite";
5          if ( r is a prime number ) then
6              if ( nⁱ mod r ≠ 1 for all i, 1 ≤ i ≤ 4⌈log n⌉² ) then
7                  break;
8          r ← r + 1 ;
9      if ( r = n ) then return  "prime";
10     for a from 1 to 2⌈√r⌉ · ⌈log n⌉ do
11         if (in ℤₙ[X]) (X + a)ⁿ mod (Xʳ − 1) ≠ Xⁿ ᵐᵒᵈ ʳ + a then
12             return  "composite";
13     return  "prime";
```

8.3 The Running Time

The analysis of the running time comes in two parts. In Sect. 8.3.1, using routine arguments, the time for the single operations of the algorithm is analyzed. The second and central part (Sect. 8.3.2) consists in proving that the **while** loop (lines 3–8) stops after $O((\log n)^c)$ cycles with r containing a number r of size $O((\log n)^c)$.

8.3.1 Overall Analysis

Time for Arithmetic Operations. Numbers are represented in binary. All numbers that occur in the execution of the algorithm are bounded by n^2, so they have bit length bounded by $2\log n$. Every arithmetic operation on such numbers can be carried out by $O((\log n)^2)$ bit operations in the naive way, and in $O^{\sim}(\log n)$ bit operations if more sophisticated methods are used (see Sect. 2.3). — In the sequel, we concentrate on bounding the *number* of arithmetic operations to be carried out by the algorithm. This number is then to be multiplied by the bit operation cost for a single arithmetic operation to obtain the overall cost in the bit model.

Time for the Perfect Power Test. The test in line 1 of the algorithm is carried out, for example, by Algorithm 2.3.5, which needs no more than $O((\log n)^2 \log \log n)$ arithmetic operations. Incidentally, the number $\lceil \log n \rceil$ may be calculated in $O(\log n)$ arithmetic operations (by repeated halving).

Time for Testing r. The loop in lines 3–8 treats the numbers $r = 2, 3, \ldots$ and tests them for several properties. Let $\rho(n)$ be the maximal r for which this loop is executed on input n. We will see in Sect. 8.3.2 that $\rho(n) = O((\log n)^c)$ for some constant c. For the time being, we bound the time needed for this loop in terms of $\rho(n)$. The test whether r divides n requires one division for each r; hence the total cost for line 4 is $O(\rho(n))$. For the decision in line 5 whether r is a prime we carry along a table of all prime numbers up to $2^{\lceil \log r \rceil}$, built up, for example, by a version of the Sieve of Eratosthenes, Algorithm 3.5.4. As soon as r reaches the value $2^i + 1$ for some i, the table of size 2^{i+1} is built, at a cost of $O(i \cdot 2^i)$, as noted in the discussion of Algorithm 3.5.4 and the subsequent estimate (3.5.9). Since, by (A.2.3) in the appendix,

$$\sum_{1 \leq i \leq \lceil \log(\rho(n)) \rceil} i \cdot 2^i \leq \log(\rho(n)) \cdot 2^{\log(\rho(n))+2} + 2 = O(\log(\rho(n)) \cdot \rho(n)),$$

the total number of arithmetic operations for maintaining and updating this table is $O(\rho(n) \log(\rho(n)))$. For the test in line 6 we calculate $n^i \bmod r$, for $i = 1, 2, \ldots, 4\lceil \log n \rceil^2$, to see whether for some of these i we get the result 1. The number of multiplications modulo r is $O((\log n)^2)$ for one r, and $O((\log n)^2 \cdot \rho(n))$ altogether.

Time for the Polynomial Operations. If the loop in lines 3–8 is left with the **break** statement in line 7, then for the number r thus determined lines 10–12 are carried out. We remark that it is trivial to find $\lceil \sqrt{r} \rceil$ in time $O(r)$. For each a, $1 \le a \le 2\lceil \sqrt{r} \rceil \cdot \lceil \log n \rceil$, the polynomial $(X + a)^n \bmod (X^r - 1)$ is calculated in $\mathbb{Z}_n[X]$, and compared with $X^{n \bmod r} + a$. From Proposition 4.3.8 we know that to calculate $(X + a)^n$ in the ring $R = \mathbb{Z}_n[X]/(X^r - 1)$ takes $O(\log n)$ ring multiplications. Since reduction modulo $X^r - 1$ is trivial (just replace X^s by X^{s-r} whenever an exponent s, $r \le s < 2r - 1$ appears), a multiplication in $\mathbb{Z}_n[X]/(X^r - 1)$ amounts to a multiplication of polynomials of degree smaller than r in the ring $\mathbb{Z}_n[X]$ and a polynomial addition. With the naive algorithm for polynomial multiplication (Definition 7.1.2) this takes $O(r^2)$ multiplications and additions of elements of $\mathbb{Z}_n$. More sophisticated polynomial multiplication algorithms carry out $O(r(\log r)(\log \log r)) = O^\sim(r)$ such multiplications (see Remark 7.1.8), such that the overall cost for calculating $(X + a)^n$ in R is $O((\log n) \cdot r^2)$ (naive) or $O^\sim((\log n) \cdot r)$ (best known algorithms). For all a taken together we can bound the number of arithmetic operations by

$$O(\sqrt{\rho(n)}(\log n) \cdot \rho(n)^2 \log n) = O(\rho(n)^{5/2}(\log n)^2) \tag{8.3.4}$$

(naive algorithms) or

$$O^\sim(\sqrt{\rho(n)}(\log n) \cdot \rho(n) \log n) = O^\sim(\rho(n)^{3/2}(\log n)^2) \tag{8.3.5}$$

(using faster algorithms). Comparing this bound with the bounds obtained for the parts of the algorithm up to line 9, we see that the time for the last loop dominates the rest whatever $\rho(n)$ may be.

Total Time. Once we have shown that $\rho(n) = O((\log n)^5)$ (which will be done in the next section), it will follow from (8.3.4) and (8.3.5) that Algorithm 8.2.1 (with naive operations) carries out $O((\log n)^{14.5})$ arithmetic operations on numbers smaller than n^2; in terms of bit operations this amounts to $O((\log n)^{16.5})$. This is the desired polynomial cost bound. If we use the more sophisticated basic algorithms, the bound on the number of arithmetic operations can be lowered to $O^\sim((\log n)^{9.5})$; the number of bit operations is bounded by $O^\sim((\log n)^{10.5})$.

8.3.2 Bound for the Smallest Witness r

This section contains the central part of the time analysis. We show that the **while** loop in lines 3–8 of the algorithm terminates after at most $20\lceil \log n \rceil^5$ iterations. For small n, we have $n < 20\lceil \log n \rceil^5$, so the loop might stop because of the test in line 3. For larger n, this means that either the loop stops because some divisor r of n has been found, or because some prime number r has been found that satisfies $\mathrm{ord}_r(n) > 4\lceil \log n \rceil^2$ — which in the loop of lines 10–12 will turn out to be a witness for n being prime or composite.

Lemma 8.3.1. *For all $n \geq 2$, there exists a prime number $r \leq 20\lceil \log n \rceil^5$ such that $r \mid n$ or $(r \nmid n$ and$)$ $\mathrm{ord}_r(n) > 4\lceil \log n \rceil^2$.*

Proof. For small n the assertion is trivially true, so we may assume that $n \geq 4$. We abbreviate $\lceil \log n \rceil$ by L, and let

$$\Pi = \prod_{1 \leq i \leq 4L^2} (n^i - 1).$$

Clearly,

$$\Pi < n^{1+2+\cdots+4L^2} = n^{8L^4+2L^2} < 2^{(\log n) \cdot 10L^4} \leq 2^{10L^5}.$$

By Proposition 3.6.9, we have

$$\prod_{r \leq 20L^5, \, r \text{ prime}} r \; > \; 2^{10L^5} > \Pi.$$

By Corollary 3.5.10, this means that there is some prime number $r \leq 20L^5$ that does not divide Π, and hence does not divide any one of the factors $n^i - 1$, $1 \leq i \leq 4L^2$. Now if r divides n, we are done. Otherwise, $\mathrm{ord}_r(n)$ is defined and larger than $4L^2 = 4\lceil \log n \rceil^2$, because $n^i \not\equiv 1 \pmod{r}$ for $1 \leq i \leq 4L^2$. $\qquad \square$

8.3.3 Improvements of the Complexity Bound

In this section, we discuss improvements of the running time bound implied by deeper results from number theory and by a number-theoretical conjecture. (This section may be skipped, since it is not used later.)

In the original version [3] of their paper, Agrawal, Kayal, and Saxena used the following deep theorem from analytic number theory. For a number n, let $P(n)$ denote the largest prime factor of n.

Theorem 8.3.2 (Fouvry [19] and Baker/Harman [9]). *There is a constant $c_0 > 0$ and some x_0 such that for all $x \geq x_0$ we have*

$$|\{p \leq x \mid p \text{ is prime and } P(p-1) \geq x^{2/3}\}| \geq c_0 \cdot \frac{x}{\ln x}.$$

One should note that there can be at most one prime factor q of $p - 1$ that satisfies $q \geq x^{2/3}$. In view of the Prime Number Theorem 3.6.2, Fouvry's theorem says that among all prime numbers p up to x at least a constant fraction will have the property that $p - 1$ has a prime factor q that exceeds $x^{2/3}$.

Using Fouvry's theorem we can find a tighter bound for the size of the smallest r that is suitable in Algorithm 8.2.1. (The proof of the following lemma is a variation of the proof of Lemma 8.3.1.)

Lemma 8.3.3. *For all sufficiently large n there exists a prime number $r \leq 8\lceil \log n \rceil^3 (\log \log n)^3$ such that $r \mid n$ or $(r \nmid n$ and$)$ $\mathrm{ord}_r(n) > 4\lceil \log n \rceil^2$.*

Proof. We let

$$x = 8\lceil \log n \rceil^3 (\log \log n)^3,$$

and note first that $x^{2/3} = 4\lceil \log n \rceil^2 (\log \log n)^2$. Fouvry's Theorem 8.3.2 tells us that the set

$$A = \{r \leq x \mid r \text{ is prime and } P(r-1) \geq x^{2/3}\}$$

has cardinality $\Omega(x/\log x) = \Omega((\log n)^3 (\log \log n)^2)$, which means that for n sufficiently large we have

$$|A| > c(\log n)^3 (\log \log n)^2, \tag{8.3.6}$$

for some constant $c > 0$. Now consider

$$\Pi = \prod_{1 \leq i \leq \lfloor x^{1/3} \rfloor} (n^i - 1).$$

It is not hard to see that $n^{x^{2/3}/3} < \Pi < n^{x^{2/3}}$, and hence that $\log(\Pi) = \Theta(x^{2/3} \log n) = \Theta((\log n)^3 (\log \log n)^2)$. If k is the number of distinct prime factors of Π, then $\Pi > 2^k \cdot k! > (2k/e)^k$, by Lemma 3.6.8. By an argument similar to that following Lemma 3.6.8, this implies that

$$k = O\left(\frac{\log(\Pi)}{\log \log(\Pi)} \right) = O\left(\frac{(\log n)^3 (\log \log n)^2}{\log \log n} \right) = O((\log n)^3 \log \log n). \tag{8.3.7}$$

Comparing the bounds in (8.3.7) and in (8.3.6) we see that for n sufficiently large we may find an r in A that does not divide Π.

Claim: $r \mid n$ or $(r \nmid n$ and$)$ $\mathrm{ord}_r(n) > 4\lceil \log n \rceil^2$.

Proof of Claim: We may assume that $r \nmid n$. Then $\mathrm{ord}_r(n)$ is defined. Since $r \in A$, we may write $r - 1 = q \cdot m$ for some prime number $q \geq x^{2/3}$. Now, by Propositions 4.1.9 and 4.2.7 the order $\mathrm{ord}_r(n)$ is a divisor of $r - 1$. Since $r \nmid \Pi$, we have that $n^i \not\equiv 1 \pmod{r}$ for $1 \leq i \leq x^{1/3}$. This means that $\mathrm{ord}_r(n) > x^{1/3} \geq (r-1)/q = m$, and hence q must divide $\mathrm{ord}_r(n)$. This implies that $\mathrm{ord}_r(n) \geq q \geq 4\lceil \log n \rceil^2 (\log \log n)^2 > 4\lceil \log n \rceil^2$, as claimed. $\square$

Taking Lemma 8.3.3 together with the discussion of the running times given in (8.3.5) at the end of Sect. 8.3.1, we obtain a bound of

$$O^\sim(\rho(n)^{3/2} (\log n)^2) = O^\sim((\log n)^{6.5}) \tag{8.3.8}$$

for the number of arithmetic operations and

$$O^\sim((\log n)^{7.5}) \tag{8.3.9}$$

for the number of bit operations. This constitutes the presently best proven complexity bound for Algorithm 8.2.1.

Finally, we briefly consider a number-theoretical conjecture, which, if true, would lower the time bounds even further. We say that a prime number $q \geq 3$ is a **Sophie Germain prime** if $2q + 1$ is a prime number as well. For example, 5, 11, 23, and 29 are Sophie Germain primes. It is conjectured that for x sufficiently large the number of Sophie Germain primes not larger than x is at least $c \cdot x/(\log x)^2$, for a suitable constant $c > 0$. If this conjecture holds, an argument quite similar to that in the proof of Lemma 8.3.3 shows that the smallest r that divides n or satisfies $\mathrm{ord}_r(n) > 4\lceil \log n \rceil^2$ is not larger than $O((\log n)^2(\log\log n)^2) = O^\sim((\log n)^2)$. In combination with the time bound from (8.3.5) we would obtain a bound of $O^\sim((\log n)^5)$ for the number of arithmetic operations and of $O^\sim((\log n)^6)$ for the number of bit operations carried out by Algorithm 8.2.1.

8.4 The Main Theorem and the Correctness Proof

The essence of the correctness proof of Algorithm 8.2.1 is given in the following theorem, which is a variant of theorems formulated and proved by D.G. Bernstein [10] on the basis of two versions [3, 4] of the paper by Agrawal, Kayal, and Saxena. In this section we use the theorem to conclude that Algorithm 8.2.1 correctly identifies prime numbers. Its proof is given in Sect. 8.5.

Theorem 8.4.1 (Main Theorem). *Assume n and r are integers so that*

(α) $n \geq 3$;
(β) $r < n$ *is a prime number;*
(γ) $a \nmid n$ *for* $2 \leq a \leq r$;
(δ) $\mathrm{ord}_r(n) > 4(\log n)^2$;
(ε) $(X + a)^n \equiv X^n + a \pmod{X^r - 1}$, *in* $\mathbb{Z}_n[X]$, *for* $1 \leq a \leq 2\sqrt{r}\log n$.

Then n is a power of a prime.

Theorem 8.4.2. *Assume Algorithm 8.2.1 is run on input $n \geq 2$. Then the output is "prime" if and only if n is a prime number.*

Proof. "$\Leftarrow$": Assume n is a prime number. Since n is not a perfect power a^b for any $b > 1$, the condition in line 1 is not satisfied. The content r of $\mathbf{r}$ is always smaller than n, hence the test in line 4 always yields that r does not divide n. If the loop in lines 3–8 is left because the content r of $\mathbf{r}$ has reached n, then in line 9 the output "*prime*" is produced. Now assume that the loop is left via the **break** statement in line 7. Let r be the content of $\mathbf{r}$ at this point. Then $r > \mathrm{ord}_r(n) > 4\lceil \log n \rceil^2 \geq 4(\log n)^2$, hence $\sqrt{r} > 2\log n$ and

$$n > r > 2\sqrt{r}\log n.$$

Because of Lemma 8.1.1 the test in line 11 will yield equality for all a with $a \leq 2\sqrt{r} \log n$ (which is smaller than n). Hence line 13 is reached and the output "*prime*" is produced.

"$\Rightarrow$": Assume Algorithm 8.2.1 outputs "*prime*" (in line 9 or in line 13).

Case 1: The loop in lines 3–8 runs until the variable **r** has attained the value n, and output "*prime*" is produced in line 9. In line 4 every r, $2 \leq r < n$, has been tested negatively for dividing n, hence n is a prime number.

Case 2: The loop in lines 3–8 is left via the **break** statement in line 7. Let $r < n$ be the content of **r** at this point. We check that n and r satisfy the conditions (α)–(ε) in Theorem 8.4.1:

(α) trivial;
(β) $r < n$ is a prime number (tested in line 5);
(γ) $a \nmid n$ for all $a \in \{2, \ldots, r\}$ (tested in the previous executions of the loop, line 4);
(δ) $\mathrm{ord}_r(n) > 4(\log n)^2$ for the order of n modulo r (tested in line 6);
(ε) $(X + a)^n \equiv (X^n + a) \pmod{X^r - 1}$, in $\mathbb{Z}_n[X]$, for $1 \leq a \leq 2\sqrt{r} \log n$ (tested in the loop in lines 10–12).

Thus, Theorem 8.4.1 applies and we conclude that $n = p^i$ for a prime number p and some $i \geq 1$. However, n cannot be a perfect power of any number because the test carried out in line 1 must have given a negative result. Hence $i = 1$, and n is a prime number. $\qquad\square$

Apart from this formal correctness proof it is also helpful to visualize the possible paths on which the algorithm reaches a result, for sufficiently large n. If the input n is a perfect power of some number, this is detected in line 1. So suppose this is not the case. In view of Lemma 8.3.1, for $n > 20\lceil \log n \rceil^5$, it is impossible that the loop in lines 3–8 runs "unsuccessfully", i.e., until **r** contains n. Instead, either a number $r \leq 20\lceil \log n \rceil^5$ that divides n is found (a witness to the compositeness of n), or a prime number $r \leq 20\lceil \log n \rceil^5$ is found so that conditions (γ) and (δ) of Theorem 8.4.1 are satisfied. Now the loop in lines 10–12 can have two different outcomes. If some a is found so that $(X + a)^n \not\equiv X^n + a \pmod{X^r - 1}$ in $\mathbb{Z}_n[X]$, then r and a together form a certificate for n being composite, by Lemma 8.1.1. If all $a \leq 2\lceil\sqrt{r}\rceil \cdot \lceil \log n \rceil$ satisfy $(X + a)^n \equiv X^n + a \pmod{X^r - 1}$ in $\mathbb{Z}_n[X]$, then n is a prime power by Theorem 8.4.1, so together with the negative outcome of the test in line 1 this constitutes a proof for the fact that n is a prime number.

8.5 Proof of the Main Theorem

This section is devoted to the proof of Theorem 8.4.1. Assume that n and r satisfy conditions (α)–(ε). Let p be an arbitrary prime divisor of n. If $p = n$, there is nothing to prove, hence we assume $p < n$, and hence $p \leq \frac{1}{2}n$. Our aim is to show that n is a power of p.

8.5.1 Preliminary Observations

The structures we will mainly work with are the field $\mathbb{Z}_p$ and the polynomial ring $\mathbb{Z}_p[X]$. In the polynomial ring we often calculate modulo some polynomial.

Note that Algorithm 8.2.1 does not use any knowledge about p, nor do we need to be able to carry out calculations in $\mathbb{Z}_p$ or $\mathbb{Z}_p[X]$ efficiently.

We abbreviate the bound occurring in condition (ε):

$$\ell = \lfloor 2\sqrt{r} \log n \rfloor, \tag{8.5.10}$$

and make some simple observations.

Lemma 8.5.1. (a) $p > r$, and r does not divide n.
(b) $r > \ell$.
(c) $1 \le a' - a < p$ for $1 \le a < a' \le \ell$.

Proof. (a) Both claims are immediate from condition (γ).
(b) Because of (δ) and the definition of $\mathrm{ord}_r(n)$, we have $r > \mathrm{ord}_r(n) > 4(\log n)^2$. This implies $\sqrt{r} > 2 \log n$, and hence $r > 2\sqrt{r} \log n \ge \ell$.
(c) Immediate from (a) and (b).   □

8.5.2 Powers of Products of Linear Terms

We are interested in the linear polynomials

$$X + a, \; 1 \le a \le \ell,$$

and products of such terms, with repetition, in $\mathbb{Z}_p[X]$:

$$P = \left\{ \prod_{1 \le a \le \ell} (X + a)^{\beta_a} \,\middle|\, \beta_a \ge 0 \text{ for } 1 \le a \le \ell \right\} \subseteq \mathbb{Z}_p[X]. \tag{8.5.11}$$

Typical examples for elements of P are 1 (all β_a are 0), $X + 1$, $X + 2$, $(X + 2)^5$, $(X + 1) \cdot (X + 3)^4 \cdot (X + 4)^3$, and so on. The purpose of this section is to establish a special property the polynomials in P have if taken to the $n^i p^j$th power, $i, j \ge 0$, if we calculate modulo $X^r - 1$. The final result is given in Lemma 8.5.6 at the end of the section. — We start with a simple consequence of condition (ε) in Theorem 8.4.1.

Lemma 8.5.2. $(X + a)^n \equiv X^n + a \pmod{X^r - 1}$, in $\mathbb{Z}_p[X]$, for $1 \le a \le \ell$.

Proof. Consider a fixed a. By condition (ε) we have $(X + a)^n \equiv X^n + a \pmod{X^r - 1}$ in $\mathbb{Z}_n[X]$. This means that there are polynomials $f, g \in \mathbb{Z}[X]$ with

$$(X + a)^n - (X^n + a) = (X^r - 1) \cdot f + n \cdot g.$$

Now p divides n, so we have $n \cdot g = p \cdot \hat{g}$ for $\hat{g} = (n/p) \cdot g$, hence $(X + a)^n \equiv (X^n + a) \pmod{X^r - 1}$ in $\mathbb{Z}_p[X]$.   □

We already know that a similar relation holds for taking the pth power.

Lemma 8.5.3. $(X + a)^p \equiv X^p + a \pmod{X^r - 1}$, *in* $\mathbb{Z}_p[X]$, *for all* $a \in \mathbb{Z}_p$.

Proof. From Proposition 7.1.15(b) we know that even $(X + a)^p = X^p + a$ in $\mathbb{Z}_p[X]$. □

In this section, we use $I(u, f)$ as an abbreviation for

$$u \geq 1 \text{ and } f \in \mathbb{Z}_p[X] \text{ and } f^u \equiv f(X^u) \pmod{X^r - 1} \text{ in } \mathbb{Z}_p[X].$$

(In [4] this relation is abbreviated as "u is ***introspective*** for f".) The last two lemmas say that $I(u, f)$ holds for $u = n$ and $u = p$ and all linear terms $X + a$, $1 \leq a \leq \ell$. We next note rules for extending this property to products in the exponent and products of polynomials.

Lemma 8.5.4. *If* $I(u, f)$ *and* $I(v, f)$, *then* $I(uv, f)$.

Proof. Since $I(v, f)$ holds, we have $f^v \equiv f(X^v) \pmod{X^r - 1}$. Applying Lemma 7.2.5(c) we conclude that

$$f^{uv} = (f^v)^u \equiv (f(X^v))^u \pmod{X^r - 1}. \tag{8.5.12}$$

Next, we apply Proposition 7.1.13(c) repeatedly to see that

$$(f(X^v))^u = \underbrace{f(X^v) \cdots f(X^v)}_{u \text{ factors}} = (\underbrace{f \cdots f}_{u \text{ factors}})(X^v) = (f^u)(X^v). \tag{8.5.13}$$

Finally, by $I(u, f)$ we may write

$$f^u - f(X^u) = (X^r - 1) \cdot g$$

for some polynomial $g \in \mathbb{Z}_p[X]$. If we substitute X^v for X, the identity remains valid, which means that

$$(f^u)(X^v) - f((X^v)^u) = ((X^v)^r - 1)g(X^v) = (X^{rv} - 1) \cdot \hat{g},$$

for $\hat{g} = g(X^v)$. We already noted (see (7.6.4)) that $X^r - 1$ is a divisor of $X^{rs} - 1$ for all $s \geq 1$, hence we can write $X^{rv} - 1 = (X^r - 1) \cdot \hat{h}$ for some $\hat{h} \in \mathbb{Z}_p[X]$. Thus

$$(f^u)(X^v) - f((X^v)^u) = (X^r - 1) \cdot \hat{h} \cdot \hat{g},$$

hence

$$(f^u)(X^v) \equiv f(X^{uv}) \pmod{X^r - 1}. \tag{8.5.14}$$

By transitivity of the congruence relation $\equiv$ we conclude from (8.5.12), (8.5.13), and (8.5.14) that

$$f^{uv} \equiv f(X^{uv}) \pmod{X^r - 1},$$

as desired. □

Lemma 8.5.5. *If $I(u, f)$ and $I(u, g)$, then $I(u, fg)$.*

Proof. We apply the hypothesis, exponentiation rules in $\mathbb{Z}_p[X]$ and $\mathbb{Z}_p$, and Lemma 7.2.5(b) to see that

$$(fg)^u = f^u \cdot g^u \equiv f(X^u) \cdot g(X^u) = (fg)(X^u) \quad (\text{mod } X^r - 1). \qquad \square$$

Lemmas 8.5.2 – 8.5.5 taken together imply that $I(u, f)$ holds for f an arbitrary product of linear terms $X + a$, $1 \leq a \leq \ell$, i.e., $f \in P$, and u an arbitrary product of n's and p's. This set of exponents is central for the considerations to follow, so we give it a name as well:

$$U = \{n^i p^j \mid i, j \geq 0\}. \tag{8.5.15}$$

The overall result of this section can now be stated as follows:

Lemma 8.5.6. *For $f \in P$ and $u \in U$ we have (in $\mathbb{Z}_p[X]$):*

$$f^u \equiv f(X^u) \quad (\text{mod } X^r - 1). \qquad \square$$

8.5.3 A Field F and a Large Subgroup G of F^*

By Proposition 7.6.4 there is some monic irreducible polynomial $h \in \mathbb{Z}_p[X]$ of degree $d = \text{ord}_r(p)$ that divides $X^{r-1} + \cdots + X + 1$ and hence $X^r - 1$. We keep this polynomial h fixed from here on, and turn our attention to the structure $F = \mathbb{Z}_p[X]/(h)$, which is a field of size p^d by Theorem 7.4.5.

Some remarks are in place. As with p and $\mathbb{Z}_p[X]$, Algorithm 8.2.1 does not refer to h at all; the existence of h is only used for the analysis. Thus, it is not necessary that operations in F can be carried out efficiently. Further, we should stress that as yet there are no restrictions we can establish on the degree d of h. Although we assume in (δ) that $\text{ord}_r(n)$ is not too small, it might even be the case that $d = \text{ord}_r(p) = 1$. (Example: For $r = 101$, $p = 607 \equiv 1 \pmod{r}$, $n = 16389 = 27 \cdot 607 \equiv 27 \pmod{r}$, the value $\text{ord}_r(n) = 100$ is as large as possible, but nonetheless $\text{ord}_r(p) = 1$.) Only later we will see that in the situation of the theorem it is not possible that $\deg(h) = 1$.

At the center of attention from here on is the subset of F that is obtained by taking the elements of P modulo h; more precisely, let

$$G = \{\, f \bmod h \mid f \in P \,\} = \Big\{ \prod_{1 \leq a \leq \ell} (X + a)^{\beta_a} \bmod h \ \Big| \ \beta_a \geq 0 \text{ for } 1 \leq a \leq \ell \Big\}. \tag{8.5.16}$$

(see (8.5.11)). — We first note that G actually is a subset of F^*.

Lemma 8.5.7. *The linear polynomials $X + a$, $1 \leq a \leq \ell$, are different in $\mathbb{Z}_p[X]$ and in F, and they satisfy $X + a \bmod h \neq 0$.*

Proof. Because of Lemma 8.5.1, the difference $(X+a')-(X+a) = a'-a$ is a nonzero element of $\mathbb{Z}_p$ and of $\mathbb{Z}_p[X]$ for $1 \le a < a' \le \ell$. Hence $X+1, \ldots, X+\ell$ are different in $\mathbb{Z}_p[X]$ and in F. Now assume for a contradiction that h divides $X + a$ for one of these a's. Since h is monic and nonconstant, we must have $h = X + a$. As noted in Proposition 7.4.6 and in Remark 7.6.3 in connection with Proposition 7.6.2, this means that $F = \mathbb{Z}_p$ and that $\zeta = -a$ is a primitive rth root of unity in $\mathbb{Z}_p$. By Lemma 8.5.2 we have $(X+a)^n \equiv X^n+a$ (mod $X^r - 1$) in $\mathbb{Z}_p[X]$; i.e., we can write

$$(X - \zeta)^n = X^n - \zeta + q \cdot (X^r - 1), \tag{8.5.17}$$

for some $q \in \mathbb{Z}_p[X]$. If we now substitute $\zeta \in \mathbb{Z}_p$ in (8.5.17) and use that $\zeta^r - 1 = 0$, we see that $\zeta^n = \zeta$, or $\zeta^{n-1} = 1$, in $\mathbb{Z}_p$. Since ζ has order r in $F^* = \mathbb{Z}_p^*$, this implies that r divides $n - 1$, that is, that $n \equiv 1 \pmod{r}$, or $\mathrm{ord}_r(n) = 1$. This contradicts condition (δ) in Theorem 8.4.1. Hence h does not divide $X + a$, for $1 \le a \le \ell$. □

Lemma 8.5.8. *G is a subgroup of F^*.*

Proof. G is the set of arbitrary products (in F) of factors $(X + a) \bmod h$. The previous lemma entails that none of these factors is 0, hence $0 \notin G$. We apply Lemma 4.1.6 to see that $G \subseteq F^*$ indeed forms a group: F^* is finite, (i) $1 = \prod_{1 \le a \le \ell}(X + a)^0$ is in G, and (ii) the product of any two elements of G is again in G. □

From Proposition 7.6.2 we know that in F the element

$$\zeta = X \bmod h$$

is a root of h and a primitive rth root of unity. The following lemma notes a crucial property that ζ has with respect to the elements of G and U.

Lemma 8.5.9 (Key Lemma). *Let $g \in G$, where $g = f \bmod h$ for $f \in P$. Then in F we have*

$$g^u = f(\zeta^u), \text{ for all } u \in U.$$

Since the rest of the correctness proof hinges on this lemma, it is important to understand clearly what the equation says: g^u is the power of $g = f \bmod h$ taken in the field F, while $f(\zeta^u)$ results from substituting the power $\zeta^u \in F$ of ζ into the polynomial $f \in \mathbb{Z}_p[X]$, evaluating in F.

Proof. Clearly, in $\mathbb{Z}_p[X]$ we have

$$g^u \equiv f^u \pmod{h}. \tag{8.5.18}$$

By Lemma 8.5.6 we know that

$$f^u \equiv f(X^u) \pmod{X^r - 1}. \tag{8.5.19}$$

Since h divides $X^r - 1$, we get (by Lemma 7.2.6) that

$$f^u \equiv f(X^u) \quad (\text{mod } h). \tag{8.5.20}$$

By definition of ζ we have $X \equiv \zeta$ (mod h). Taking uth powers, we get (by Proposition 7.2.5(b)) $X^u \equiv (\zeta^u \bmod h)$ (mod h). Substituting both terms into f we obtain, by Lemma 7.2.5(c), that

$$f(X^u) \equiv f(\zeta^u \bmod h) \quad (\text{mod } h). \tag{8.5.21}$$

Now combining (8.5.18), (8.5.20), and (8.5.21) yields

$$g^u \bmod h = f^u \bmod h = f(X^u) \bmod h = f(\zeta^u \bmod h) \bmod h.$$

Since $g^u \bmod h$ is g^u in F, and $\zeta^u \bmod h$ is ζ^u in F, and $f(\zeta^u \bmod h) \bmod h$ is the result of substituting $\zeta^u \in F$ into f, this is the assertion of the lemma.
$\qquad\square$

Because of the Key Lemma 8.5.8, the following set of powers of ζ in F seems to be interesting. Let

$$T = \{\zeta^u \mid u \in U\} \quad, \text{ and } \quad t = |T|. \tag{8.5.22}$$

There is no reason to assume that T is closed under multiplication; it is just a set. — We note that condition (δ) in Theorem 8.4.1 enforces that T is not too small.

Lemma 8.5.10.

$$r > t > 4(\log n)^2.$$

Proof. *Upper bound*: Recall that $T \subseteq \langle \zeta \rangle$ and that $\langle \zeta \rangle = \{1, \zeta, \ldots, \zeta^{r-1}\}$. Now $\zeta^u \neq 1$ for all $u \in U$ (note that r does not divide $n^i p^j$ for any $i, j \geq 0$ and apply Proposition 4.2.7(b)); thus, $t = |T| \leq r - 1$.
Lower bound: By its definition, the set T contains at least all the powers ζ^{n^i}, $i = 0, 1, 2, \ldots$, in F^*. Now $\langle \zeta \rangle$ is a cyclic group of size r. Thus we may apply Proposition 4.2.7(b) to see that ζ^{n^i} and ζ^{n^k} are distinct if and only if $n^i \bmod r \neq n^k \bmod r$. This means that

$$|\{\zeta^{n^i} \mid i \geq 0\}| = |\{n^i \bmod r \mid i \geq 0\}| = \text{ord}_r(n).$$

Hence $t = |T| \geq \text{ord}_r(n)$. Using condition ($\delta$) in Theorem 8.4.1 we conclude that $t > 4(\log n)^2$.
$\qquad\square$

The set T is instrumental in cutting out a large portion of P on which the mapping $P \ni f \mapsto f \bmod h \in G$ is one-to-one; this, in turn, will enable us to establish a large lower bound on $|G|$. For the proof, we employ Key Lemma 8.5.9.

Lemma 8.5.11. *If f_1 and f_2 are distinct elements of P with $\deg(f_1)$, $\deg(f_2) < t$, then $f_1 \bmod h \neq f_2 \bmod h$.*

(Note that a conclusion like that cannot be drawn for polynomials f_1, f_2 that are not in P, since $\deg(h)$ may be much smaller than t.)

Proof. Indirect. Let $g_1 = f_1 \bmod h$ and $g_2 = f_2 \bmod h$, and assume for a contradiction that $g_1 = g_2$. Let $u \in U$ be arbitrary. By Lemma 8.5.9 we get

$$f_1(\zeta^u) = g_1^u = g_2^u = f_2(\zeta^u)$$

(calculating in F). This means that $f_1(z) = f_2(z)$ for *all* elements z of T, of which there are t many. On the other hand, $\deg(f_1)$, $\deg(f_2) < t$, by assumption. Hence, by Corollary 7.5.2, we must have $f_1 = f_2$, a contradiction.
□

Using Lemma 8.5.11, we may prove a large lower bound on the cardinality of G.

Lemma 8.5.12.

$$|G| > \frac{1}{2} n^{2\sqrt{t}}.$$

Proof. Let $\mu = \min\{\ell, t - 1\}$. Then the polynomials

$$\prod_{1 \le a \le \mu} (X + a)^{\beta_a}, \quad \beta_a \in \{0, 1\} \text{ for } 1 \le a \le \mu,$$

are all in P and have degree smaller than t. They are given explicitly as products of different sets of irreducible factors; it follows from the Unique Factorization Theorem for polynomials (Theorem 7.4.4) that they are different in $\mathbb{Z}_p[X]$, and hence in P. Hence taking them modulo h yields different elements of G, by Lemma 8.5.11. This shows that $|G| \ge 2^\mu$.
Case 1: $\mu = \ell$. — Then, by the bound $r > t$ from Lemma 8.5.10:

$$\mu = \lfloor 2\sqrt{r} \log n \rfloor > 2\sqrt{r} \log n - 1 > 2\sqrt{t} \log n - 1.$$

Case 2: $\mu = t - 1$. — Then, by the bound $t > 4(\log n)^2$ from Lemma 8.5.10:

$$\mu = t - 1 > 2\sqrt{t} \log n - 1.$$

In both cases, we obtain

$$|G| \ge 2^\mu > 2^{2\sqrt{t} \log n - 1} = \frac{1}{2} n^{2\sqrt{t}},$$

as desired.
□

Remark 8.5.13. Incidentally, combining Lemmas 8.5.10 and 8.5.12 we see that $|G| > \frac{1}{2} n^{2\sqrt{t}} > \frac{1}{2} n^{4 \log n}$. Thus, $|F| > |G| > p^{4 \log n}$, which implies that $d = \deg(h) > 4 \log n$. So $F \ne \mathbb{Z}_p$, and $\zeta = X$ after all.

8.5.4 Completing the Proof of the Main Theorem

In this section, we finish the proof of Theorem 8.4.1. We use the field F, the subgroup G of F^*, the rth root of unity ζ in F, the set U of exponents, and the set $T \subseteq \langle\zeta\rangle$ with its cardinality t from the previous section. Consider the following finite subset of U:

$$U_0 = \{n^i p^j \mid 0 \leq i, j \leq \lfloor\sqrt{t}\rfloor\} \subseteq U.$$

The definition of U_0 is designed so that the following upper bound on the size of U_0 can be proved, employing the Key Lemma 8.5.9 again, but in a way different than in the proof of Lemma 8.5.11.

Lemma 8.5.14. $|U_0| \leq t$.

Proof. Claim 1: $u < |G|$, for all $u \in U_0$.
Proof of Claim 1: We assumed that p is a proper divisor of n, so $p \leq \frac{1}{2}n$. Thus, for $i, j \leq \lfloor\sqrt{t}\rfloor$ we have

$$n^i p^j \leq \left(\frac{1}{2}n^2\right)^{\sqrt{t}} \leq \frac{1}{2}n^{2\sqrt{t}} < |G|,$$

by Lemma 8.5.12. Thus Claim 1 is proved.

Since $t = |T| = |\{\zeta^u \mid u \in U\}|$, to establish Lemma 8.5.14 it is sufficient to show that the mapping $U \ni u \mapsto \zeta^u \in T$ is one-to-one on U_0, as stated next.
Claim 2: If $u, v \in U_0$ are different, then $\zeta^u \neq \zeta^v$ in F.
Proof of Claim 2: Indirect. Assume that $u, v \in U_0$ and $\zeta^u = \zeta^v$. Let $g \in G$ be arbitrary; by the definition of G we may write $g = f \bmod h$ for some $f \in P$. Applying the Key Lemma 8.5.9 we see that (calculating in F) we have

$$g^u = f(\zeta^u) = f(\zeta^v) = g^v.$$

This can be read so as to say that g is a root (in F) of the polynomial $X^u - X^v \in \mathbb{Z}_p[X]$. Now $g \in G$ was arbitrary, so *all* elements of G are roots of $X^u - X^v$. On the other hand, $\deg(X^u - X^v) \leq \max\{u, v\} < |G|$, by Claim 1. By Theorem 7.5.1 this implies that $X^u - X^v$ is the zero polynomial. This means that $u = v$, which is the desired contradiction. Thus Claim 2 is proved, and Lemma 8.5.14 follows.   □

Now, at last, we can finish the proof of the Main Theorem 8.4.1.

Lemma 8.5.15. *n is a power of p.*

Proof. It is clear that the number of pairs (i, j), $0 \leq i, j \leq \lfloor\sqrt{t}\rfloor$, is $(\lfloor\sqrt{t}\rfloor + 1)^2 > \sqrt{t}^2 = t$. On the other hand, Lemma 8.5.14 says that $U_0 = \{n^i p^j \mid 0 \leq i, j \leq \lfloor\sqrt{t}\rfloor\}$ does not have more than t elements. By the pigeon hole principle there must be two distinct pairs $(i, j), (k, m)$ with $n^i p^j = n^k p^m$. Clearly, it is not possible that $i = k$ (otherwise $j = m$ would follow); by symmetry, we may assume that $i > k$. Consequently,

$$n^{i-k} = p^{m-j},$$

with $i - k > 0$ and $m - j > 0$. Thus, by the Fundamental Theorem of Arithmetic (Theorem 3.5.8), n cannot have any prime factors besides p, and the lemma is proved.   □

A. Appendix

A.1 Basics from Combinatorics

The *factorial function* is defined by

$$n! = 1 \cdot 2 \cdot \ldots \cdot n = \prod_{1 \leq i \leq n} i \, , \quad \text{for integers } n \geq 0.$$

As the empty product has value 1, we have $0! = 1! = 1$. Further, $2! = 2$, $3! = 6$, $4! = 24$, and so on. In combinatorics, $n!$ is known to be the number of permutations of $\{1, \ldots, n\}$, i.e., the number of ways in which n different objects can be arranged as a sequence. In calculus, one comes across factorials in connection with Taylor series, and in particular in the series for the exponential function ($e \approx 2.718$ is the base of the natural logarithm):

$$e^x = \sum_{i \geq 0} \frac{x^i}{i!} \, , \quad \text{for all real } x. \tag{A.1.1}$$

As an easy consequence of this we note that for $n \geq 1$ we have $n^n/n! < \sum_{i \geq 0} n^i/i! = e^n$ and hence the following bound:

$$n! > \left(\frac{n}{e}\right)^n. \tag{A.1.2}$$

The *binomial coefficients* are defined as follows:

$$\binom{n}{k} = \frac{n!}{k!(n-k)!} = \frac{n(n-1)\cdots(n-k+1)}{k!} \, , \tag{A.1.3}$$

for integers $n \geq 0$ and $0 \leq k \leq n$. It is useful to extend the definition to $\binom{n}{k} = 0$ for $k < 0$ and $k > n$. Although the binomial coefficients look like fractions, they are really integers. This is easily seen by considering their combinatorial interpretation:

Fact A.1.1. *Let A be an arbitrary n-element set. Then for any integer k there are exactly $\binom{n}{k}$ subsets of A with k elements.*

M. Dietzfelbinger: Primality Testing in Polynomial Time, LNCS 3000, pp. 133-142, 2004.
© Springer-Verlag Berlin Heidelberg 2004

Proof. For $k < 0$ or $k > n$, there is no such subset, and $\binom{n}{k} = 0$. Thus, assume $0 \leq k \leq n$. We consider the set S_n of all permutations of A, i.e., sequences $(a_1, \ldots, a_n)$ in which every element of A occurs exactly once. We know that S_n has exactly $n!$ elements. We now count the elements of S_n again, grouped in a particular way according to the k-element subsets. For an arbitrary k-element subset B of A, let

$$S_B = \{(a_1, \ldots, a_n) \in S_n \mid \{a_1, \ldots, a_k\} = B\} \ .$$

Since the elements of B can be arranged in $k!$ ways in the first k positions of a sequence and likewise the elements of $A - B$ can be arranged in $(n - k)!$ ways in the last $n - k$ positions, we get $|S_B| = k!(n - k)!$. Now, obviously,

$$S_n = \bigcup_{B \subseteq A, |B| = k} S_B \ ,$$

a union of disjoint sets. Thus,

$$n! = |S_n| = \sum_{B \subseteq A, |B| = k} |S_B| = |\{B \mid B \subseteq A, |B| = k\}| \cdot k!(n - k)! \ ,$$

which proves the claim. $\qquad\square$

Taking in particular $A = \{1, \ldots, n\}$, the previous fact is equivalent to saying that $\binom{n}{k}$ is the number of n-bit 0-1-strings $(a_1, \ldots, a_n)$ with exactly k 1's. Since there are 2^n many n-bit strings altogether, we observe:

$$\sum_{0 \leq k \leq n} \binom{n}{k} = 2^n \ . \tag{A.1.4}$$

We note the following important recursion formula for the binomial coefficients:

$$\binom{n}{0} = \binom{n}{n} = 1 \ , \text{ for all } n \geq 0 \ ; \tag{A.1.5}$$

$$\binom{n}{k} = \binom{n-1}{k-1} + \binom{n-1}{k} \ , \text{ for all } n \geq 1, \text{ all integers } k \ . \tag{A.1.6}$$

(Note that these formulas give another proof of the fact that $\binom{n}{k}$ is an integer.)

Formula (A.1.6) is obvious for $k \leq 0$ and $k \geq n$. For $1 \leq k \leq n - 1$ it can be verified directly from the definition, as follows:

$$\binom{n}{k} - \binom{n-1}{k-1}$$

$$= \frac{n(n-1) \cdots (n - k + 1)}{k!} - \frac{k(n-1)(n-2) \cdots (n - k + 1)}{(k-1)! \cdot k}$$

$$= \frac{(n - k) \cdot (n - 1) \cdots (n - k + 1)}{k!} = \binom{n-1}{k} \ .$$

Formulas (A.1.5) and (A.1.6) give rise to "Pascal's triangle", a pattern to generate all binomial coefficients.

$$
\begin{array}{ccccccccccccccccccccc}
&&&&&&&&&& 1 \\
&&&&&&&&& 1 && 1 \\
&&&&&&&& 1 && 2 && 1 \\
&&&&&&& 1 && 3 && 3 && 1 \\
&&&&&& 1 && 4 && 6 && 4 && 1 \\
&&&&& 1 && 5 && 10 && 10 && 5 && 1 \\
&&&& 1 && 6 && 15 && 20 && 15 && 6 && 1 \\
&&& 1 && 7 && 21 && 35 && 35 && 21 && 7 && 1 \\
&& 1 && 8 && 28 && 56 && 70 && 56 && 28 && 8 && 1 \\
& 1 && 9 && 36 && 84 && 126 && 126 && 84 && 36 && 9 && 1 \\
1 && 10 && 45 && 120 && 210 && 252 && 210 && 120 && 45 && 10 && 1
\end{array}
$$

Row n has $n+1$ entries $\binom{n}{k}$, $0 \le k \le n$. The first and last entry in each row is 1, each other entry is obtained by adding the two values that are above it (to the north-west and to the north-east).

Next we note some useful estimates involving binomial coefficients. They say that Pascal's triangle is symmetric; and that in each row the entries increase up to the center, then decrease. Finally, bounds for the central entry $\binom{2n}{n}$ in the even-numbered rows are given.

Lemma A.1.2. (a) $\binom{n}{k} = \binom{n}{n-k}$, for $0 \le k \le n$.
(b) If $1 \le k \le \frac{n}{2}$, then $\binom{n}{k-1} < \binom{n}{k}$.
(c) $2^n \le \frac{2^{2n}}{2n} \le \binom{2n}{n} < 2^{2n}$, for $n \ge 1$.

Proof. (a) Note that $\binom{n}{k} = \frac{n!}{k!(n-k)!} = \frac{n!}{(n-k)!(n-(n-k))!} = \binom{n}{n-k}$.
(b) Observe that

$$
\frac{\binom{n}{k}}{\binom{n}{k-1}} = \frac{n-k+1}{k} \ge \frac{n/2+1}{n/2} > 1.
$$

(c) The last inequality $\binom{2n}{n} < 2^{2n}$ is a direct consequence of (A.1.4). For the second inequality $\frac{2^{2n}}{2n} \le \binom{2n}{n}$ observe that by parts (a) and (b) $\binom{2n}{n}$ is maximal in the set containing $\binom{2n}{0} + \binom{2n}{2n} = 1+1 = 2$ and $\binom{2n}{i}$, $0 < i < 2n$. Hence $\binom{2n}{n}$ is at least the average of these $2n$ numbers, which is $2^{2n}/2n$ by (A.1.4). The first inequality is equivalent to $2n \le 2^n$, which is obviously true for $n \ge 1$. $\square$

The binomial coefficients are important in expressing powers of sums. Assume $(R, +, \cdot, 0, 1)$ is a commutative ring (see Definition 4.3.2). For $a \in R$ and $m \ge 0$ we write a^m as an abbreviation of $a \cdots a$ (m factors), and m_R

as an abbreviation of the sum $1 + \cdots + 1$ (m summands) in R. Then for all $a, b \in R$ and $n \geq 0$ we have:

$$(a + b)^n = \sum_{0 \leq k \leq n} \binom{n}{k}_R \cdot a^k b^{n-k} \,. \tag{A.1.7}$$

This formula is often called the **binomial theorem**. It is easy to prove, using the combinatorial interpretation of the binomial coefficients. The cases $n = 0$ and $n = 1$ are trivially true, so assume $n \geq 2$ and consider the product $(a + b) \cdots (a + b)$ with n factors. If this is expanded by "multiplying out" in R, we obtain a sum of 2^n products of n factors each, where a product contains either the a or the b from each factor $(a + b)$. By Fact A.1.1, there are exactly $\binom{n}{k}$ products in which a occurs k times and b occurs $n-k$ times. By commutativity, each such product equals $a^k b^{n-k}$. Finally, we write $a^k b^{n-k} + \cdots + a^k b^{n-k}$ as $(1 + \cdots + 1) \cdot a^k b^{n-k}$, with $\binom{n}{k}$ summands in each case.

A.2 Some Estimates

Definition A.2.1. *For $x \in \mathbb{R}$ we let $\lfloor x \rfloor$ ("floor of x") denote the largest integer k with $k \leq x$. Similarly, $\lceil x \rceil$ ("ceiling of x") denotes the smallest integer k with $k \geq x$.*

For example, $\lfloor 5.95 \rfloor = 5$ and $\lfloor 6.01 \rfloor = 6$. Clearly, $\lfloor x \rfloor$ is characterized by the fact that it is an integer and that $x - 1 < \lfloor x \rfloor \leq x$. Similarly, the characteristic inequalities for $\lceil x \rceil$ are $x \leq \lceil x \rceil < x + 1$. If $a \geq 0$ and $b > 0$ are integers, then, with the notation of Definition 3.1.9,

$$\left\lfloor \frac{a}{b} \right\rfloor = a \text{ div } b,$$

hence in particular

$$a = \left\lfloor \frac{a}{b} \right\rfloor \cdot b + (a \bmod b). \tag{A.2.8}$$

A basic property of the floor function is the following:

Lemma A.2.2. *For all real numbers $y \geq 0$ we have $\lfloor 2y \rfloor - 2\lfloor y \rfloor \in \{0, 1\}$.*

Proof. Let $\{y\} = y - \lfloor y \rfloor < 1$ be the "fractional part" of y. Then $0 \leq \{y\} < 1$. If $0 \leq \{y\} < \frac{1}{2}$, then $2\lfloor y \rfloor \leq 2y < 2\lfloor y \rfloor + 1$, hence $\lfloor 2y \rfloor = 2\lfloor y \rfloor$. If $\frac{1}{2} \leq \{y\} < 1$, then $2\lfloor y \rfloor + 1 \leq 2y < 2\lfloor y \rfloor + 2$, hence $\lfloor 2y \rfloor = 2\lfloor y \rfloor + 1$. $\square$

Next, we estimate a power sum and the harmonic sum.

Lemma A.2.3. *For all $n \geq 0$ we have*

$$\sum_{1 \leq i \leq n} i \cdot 2^i = (n - 1) \cdot 2^{n+1} + 2.$$

Proof. We use induction on n. For $n = 0$ the claim is easily checked. The induction step follows from the observation that

$$(n-1) \cdot 2^{n+1} + 2 - ((n-2) \cdot 2^n + 2) = (2(n-1) - (n-2)) \cdot 2^n = n \cdot 2^n.$$

$\square$

Lemma A.2.4. *For $H_n = \sum_{1 \leq i \leq n} \frac{1}{i}$ (the nth **harmonic number**) we have*

$$\ln n \ < \ H_n \ < \ 1 + \ln n,, \ \textit{for } n \geq 2.$$

Proof. Note that

$$\int_i^{i+1} \frac{dx}{x} < \frac{1}{i}, \text{ for } i \geq 1, \text{ and } \frac{1}{i} < \int_{i-1}^i \frac{dx}{x}, \text{ for } i \geq 2.$$

Summing the first inequality for $1 \leq i < n$, we obtain

$$\ln n = \int_1^n \frac{dx}{x} = \sum_{1 \leq i < n} \int_i^{i+1} \frac{dx}{x} < \sum_{1 \leq i < n} \frac{1}{i} = H_n - \frac{1}{n} < H_n;$$

summing the second inequality for $1 < i \leq n$ we obtain

$$H_n - 1 = \sum_{1 < i \leq n} \frac{1}{i} < \sum_{1 < i \leq n} \int_{i-1}^i \frac{dx}{x} = \int_1^n \frac{dx}{x} = \ln n.$$

$\square$

A.3 Proof of the Quadratic Reciprocity Law

In this section, we provide a full proof of Theorem 6.3.1, the quadratic reciprocity law. Also, Proposition 6.3.2 will be proved here.

A.3.1 A Lemma of Gauss

Let $p \geq 3$ be a prime number. We let

$$H_p = \{1, 2 \ldots, \tfrac{1}{2}(p-1)\}.$$

Traditionally, H_p is called the "canonical half system", since it contains exactly half the elements of $\mathbb{Z}_p^*$, and $\mathbb{Z}_p^* = H \cup \{p - i \mid i \in H\}$, as a union of disjoint sets. (Recall that $p - i$ is the additive inverse $-i$ of i in $\mathbb{Z}_p$.) For $a \in \mathbb{Z}$ with $p \nmid a$ consider the sequence

$$S_p(a) = ((a \cdot 1) \bmod p, (a \cdot 2) \bmod p, \ldots, (a \cdot \tfrac{1}{2}(p-1)) \bmod p).$$

Note that, clearly, $S_p(a) = S_p(b)$ if $a \equiv b \pmod p$. Some of the entries in $S_p(a)$ will be in H_p, some will be not.

For a with $p \nmid a$ we define:

$$k_p(a) = \text{the number of entries in } S_p(a) \text{ that belong to } \mathbb{Z}_p^* - H_p.$$

a	1	2	3	4	5	6	7	8	9	10	11	12	13	14	15	16
$2a \bmod 17$	2	4	6	8	10	12	14	16	1	3	5	7	9	11	13	15
$3a \bmod 17$	3	6	9	12	15	1	4	7	10	13	16	2	5	8	11	14
$4a \bmod 17$	4	8	12	16	3	7	11	15	2	6	10	14	1	5	9	13
$5a \bmod 17$	5	10	15	3	8	13	1	6	11	16	4	9	14	2	7	12
$6a \bmod 17$	6	12	1	7	13	2	8	14	3	9	15	4	10	16	5	11
$7a \bmod 17$	7	14	4	11	1	8	15	5	12	2	9	16	6	13	3	10
$8a \bmod 17$	8	16	7	15	6	14	5	13	4	12	3	11	2	10	1	9
$k_{17}(a)$	0	4	3	4	3	3	3	4	4	5	5	5	4	5	4	8
sign of $(-1)^{k_{17}(a)}$	+	+	−	+	−	−	−	+	+	−	−	−	+	−	+	+

Table A.1. $S_{17}(a)$ and $k_{17}(a)$, for $a = 1, \ldots, 16$

Lemma A.3.1. *If the odd prime p does not divide a, then*

$$\left(\frac{a}{p}\right) = (-1)^{k_p(a)}.$$

In words: a is a quadratic residue modulo p if and only if $k_p(a)$ is even.

As an example, consider $p = 17$. The canonical half system is $H_{17} = \{1, \ldots, 8\}$. The sequences $S_{17}(a)$, $1 \le a \le 16$, are listed in Table A.1, together with $k_{17}(a)$ and the sign of $(-1)^{k_{17}(a)}$. Assuming the lemma for a moment, we may read off from the table that the quadratic residues modulo 17 in $\mathbb{Z}_{17}^*$ are $1, 2, 4, 8, 9, 13, 15, 16$.

Proof. Let $k = k_p(a)$, let $H = H_p$, let $R \subseteq \mathbb{Z}_p^*$ the set of k entries in $S_p(a)$ that exceed $p/2$, and let T be the set of the $\frac{1}{2}(p-1) - k$ other entries.
Claim: H is the disjoint union of $\{p - r \mid r \in R\}$ and T.
Proof of Claim: Since $\mathbb{Z}_p$ is a field, the entries in $S_p(a)$ are distinct, hence $\{p - r \mid r \in R\} \subseteq H$ has k elements and T has $p/2 - k$ elements. Thus, to prove the claim, it is sufficient to show that $\{p - r \mid r \in R\} \cap T = \emptyset$. Assume for a contradiction that $p - r = t$ for some $r \in R$ and some $t \in T$. Now $r = i \cdot a \bmod p$ for some $i \in H$, and $t = j \cdot a \bmod p$ for some $j \in H$. The assumption entails

$$0 \equiv r + t \equiv i \cdot a + j \cdot a \equiv (i + j) \cdot a \pmod{p}.$$

But $p \nmid a$ and $p \nmid (i + j)$, since $0 < i + j < p$. This is the desired contradiction, and the claim is proved.

(For an example, the reader may wish to go back to Table A.1 and in the sequences $S_{17}(a)$ replace the entries r larger than 8 by $17 - r$ to see that in each case the resulting sequence is just a permutation of $(1, 2, \ldots, 8)$.)

Let $b = a \bmod p$, and calculate in $\mathbb{Z}_p$:

$$\prod_{i \in H} i = \prod_{r \in R}(-r) \cdot \prod_{t \in T} t = (-1)^k \cdot \prod_{r \in R} r \cdot \prod_{t \in T} t = (-1)^k \prod_{i \in H}(i \cdot b)$$
$$= (-1)^k \cdot b^{(p-1)/2} \cdot \prod_{i \in H} i.$$

By cancelling in $\mathbb{Z}_p^*$, we conclude $(-1)^k \cdot b^{(p-1)/2} = 1$ in $\mathbb{Z}_p$. Since both $(-1)^k$ and $b^{(p-1)/2}$ belong to $\{1, -1\}$ in $\mathbb{Z}_p$, this entails $(-1)^k \equiv b^{(p-1)/2} \pmod{p}$. By Lemma 6.1.3, we conclude that $(-1)^k \equiv \left(\frac{b}{p}\right) \bmod p \equiv \left(\frac{a}{p}\right) \bmod p$; this means that k is even if and only if $\left(\frac{a}{p}\right) = 1$. □

Using Lemma A.3.1, it is easy to determine when 2 is a quadratic residue modulo p.

Corollary A.3.2. *For $p \geq 3$ a prime number, we have:*

$$\left(\frac{2}{p}\right) = (-1)^{(p^2-1)/8}.$$

In words: 2 is a quadratic residue modulo p if and only if $p \equiv 1$ or $p \equiv 7$ (mod 8).

Proof. The number $k_p(2)$ of elements of size at least $p/2$ in the sequence $S_p(2) = (2, 4, 6, \ldots, p-1)$ is the same as the number of elements of size at least $p/4$ in $(1, 2, 3, \ldots, \frac{1}{2}(p-1))$. Since $p/4$ is not an integer, we have

$$k_p(2) = \tfrac{1}{2}(p-1) - \lfloor p/4 \rfloor.$$

Depending on the remainder $p \bmod 8$, there are four cases:
If $p = 8\ell + 1$, then $k_p(2) = 4\ell - 2\ell = 2\ell$, which is even.
If $p = 8\ell + 3$, then $k_p(2) = 4\ell + 1 - 2\ell = 2\ell + 1$, which is odd.
If $p = 8\ell + 5$, then $k_p(2) = 4\ell + 2 - (2\ell + 1) = 2\ell + 1$, which is odd.
If $p = 8\ell + 7$, then $k_p(2) = 4\ell + 3 - (2\ell + 1) = 2\ell + 2$, which is even.
The claim now follows from Lemma A.3.1. □

A.3.2 Quadratic Reciprocity for Prime Numbers

For $a \in \mathbb{Z}$, $p \nmid a$, let

$$\lambda_p(a) = \sum_{i \in H_p}(i \cdot a) \operatorname{div} p = \sum_{i \in H_p}\left\lfloor \frac{i \cdot a}{p}\right\rfloor = \sum_{i \in H_p}\frac{i \cdot a - (i \cdot a) \bmod p}{p}. \quad (A.3.9)$$

(The last equation in this definition is immediate from (A.2.8). Note that $\lambda_p(a)$ depends on a, not only on the equivalence class of a modulo p.)

Lemma A.3.3. *If $p \geq 3$ is a prime number and $a \in \mathbb{Z}$ is odd with $p \nmid a$, then*

$$\left(\frac{a}{p}\right) = (-1)^{\lambda_p(a)}.$$

In words: a is a quadratic residue modulo p if and only if $\lambda_p(a)$ is even.

Proof. Using the definition of R and T from Lemma A.3.1, and writing H for H_p again, we calculate in $\mathbb{Z}$:

$$\sum_{i \in H} i \cdot a = \sum_{i \in H} (i \cdot a - (i \cdot a) \bmod p) + \sum_{r \in R} r + \sum_{t \in T} t = \lambda_p(a) \cdot p + \sum_{r \in R} r + \sum_{t \in T} t.$$

$$\text{(A.3.10)}$$

Using the claim in the proof of Lemma A.3.1 again, we obtain

$$\sum_{i \in H} i = \sum_{r \in R} (p - r) + \sum_{t \in T} t = k_p(a) \cdot p - \sum_{r \in R} r + \sum_{t \in T} t. \qquad \text{(A.3.11)}$$

Subtracting (A.3.11) from (A.3.10) yields

$$(a - 1) \cdot \sum_{i \in H} i = (\lambda_p(a) - k_p(a)) \cdot p + 2 \cdot \sum_{r \in R} r. \qquad \text{(A.3.12)}$$

Now since a is odd, $a - 1$ is even, so (A.3.12) implies that $\lambda_p(a) - k_p(a)$ is an even number. Thus, the lemma follows by Lemma A.3.1. □

For example, we could calculate that $\lambda_{17}(10) = \lfloor \frac{10}{17} \rfloor + \lfloor \frac{20}{17} \rfloor + \lfloor \frac{30}{17} \rfloor + \lfloor \frac{40}{17} \rfloor + \lfloor \frac{50}{17} \rfloor + \lfloor \frac{60}{17} \rfloor + \lfloor \frac{70}{17} \rfloor + \lfloor \frac{80}{17} \rfloor = 0 + 1 + 1 + 2 + 2 + 3 + 4 + 4 = 19$, which is odd; using Lemma A.3.1 we conclude that $\left(\frac{10}{17}\right) = -1$. Obviously, in general it is sufficient to add the numbers $\lfloor \frac{i \cdot a}{p} \rfloor$, $1 \leq i \leq \frac{1}{2}(p - 1)$, modulo 2. However, this is a hopelessly inefficient method for calculating $\left(\frac{a}{p}\right)$. We will use Lemma A.3.3 only for proving the following theorem.

Theorem A.3.4 (Quadratic Reciprocity for Prime Numbers). *Let p and q be distinct odd prime numbers. Then*

$$\left(\frac{p}{q}\right) = (-1)^{\frac{p-1}{2} \cdot \frac{q-1}{2}} \cdot \left(\frac{q}{p}\right),$$

that means

$$\left(\frac{p}{q}\right) = \begin{cases} \left(\dfrac{q}{p}\right) & , \text{ if } p \equiv 1 \text{ or } \quad q \equiv 1 \pmod 4 \text{ , and} \\[2ex] -\left(\dfrac{q}{p}\right) & , \text{ if } p \equiv 3 \text{ and } q \equiv 3 \pmod 4. \end{cases}$$

Proof. Let

$$M = \left\{ (i,j) \,\middle|\, 1 \le i \le \frac{p-1}{2},\ 1 \le j \le \frac{q-1}{2} \right\}. \tag{A.3.13}$$

Now define

$$M_1 = \{(i,j) \in M \mid j \cdot p < i \cdot q\} \text{, and}$$
$$M_2 = \{(i,j) \in M \mid i \cdot q < j \cdot p\}.$$

Note that there cannot be a pair $(i,j) \in M$ that satisfies $i \cdot q = j \cdot p$, since this would mean that p divides i, which is smaller than p. Thus, M_1 and M_2 split M into two disjoint subsets, and we get

$$|M_1| + |M_2| = |M| = \frac{p-1}{2} \cdot \frac{q-1}{2}. \tag{A.3.14}$$

Now for each fixed $i \le \frac{p-1}{2}$ the number of pairs $(i,j) \in M_1$ is $\lfloor i \cdot q/p \rfloor$. Hence

$$|M_1| = \sum_{1 \le i \le (p-1)/2} \lfloor i \cdot q/p \rfloor = \lambda_p(q).$$

Similarly, $|M_2| = \lambda_q(p)$. Thus, from (A.3.14) we get

$$\frac{p-1}{2} \cdot \frac{q-1}{2} = \lambda_p(q) + \lambda_q(p).$$

Now Lemma A.3.3 entails

$$(-1)^{\frac{p-1}{2} \cdot \frac{q-1}{2}} = (-1)^{\lambda_p(q)+\lambda_q(p)} = \left(\frac{p}{q}\right) \cdot \left(\frac{q}{p}\right),$$

which is the assertion of the theorem.   ⊓

A.3.3 Quadratic Reciprocity for Odd Integers

In this section, we prove Theorem 6.3.1 and Proposition 6.3.2.

Proof of Theorem 6.3.1. We show: If $n \ge 3$ and $m \ge 3$ are odd integers, then

$$\left(\frac{m}{n}\right) = (-1)^{\frac{n-1}{2} \cdot \frac{m-1}{2}} \cdot \left(\frac{n}{m}\right).$$

We consider the prime factorizations $n = p_1 \cdots p_r$ and $m = q_1 \cdots q_s$, and prove the claim by induction on $r + s$. If m and n are not relatively prime, both $\left(\frac{n}{m}\right)$ and $\left(\frac{m}{n}\right)$ are 0, and there is nothing to show. Thus, we assume from here on that $\gcd(n,m) = 1$.

Basis: $r + s = 2$, i.e., n and m are distinct prime numbers. — Then the claim is just Theorem A.3.4.

Induction step: Assume $r + s \geq 3$, and the claim is true for all n', m' that together have fewer than $r + s$ prime factors. By symmetry, we may assume that n is not a prime number. We write $n = k\ell$ for numbers k, $\ell \geq 3$. By the induction hypothesis, we have

$$\left(\frac{m}{k}\right) \cdot \left(\frac{k}{m}\right) = (-1)^{\frac{k-1}{2} \cdot \frac{m-1}{2}} \quad \text{and} \quad \left(\frac{m}{\ell}\right) \cdot \left(\frac{\ell}{m}\right) = (-1)^{\frac{\ell-1}{2} \cdot \frac{m-1}{2}}.$$

$$(A.3.15)$$

Using multiplicativity in both upper and lower positions (Lemma 6.2.2(a) and (c)) we get by multiplying both equations:

$$\left(\frac{m}{n}\right) \cdot \left(\frac{n}{m}\right) = (-1)^{\frac{k-1}{2} \cdot \frac{m-1}{2} + \frac{\ell-1}{2} \cdot \frac{m-1}{2}} = \left((-1)^{\frac{k-1}{2} + \frac{\ell-1}{2}}\right)^{\frac{m-1}{2}}.$$

$$(A.3.16)$$

Now $\frac{(k-1)(\ell-1)}{2}$ is an even number, hence

$$\frac{n-1}{2} = \frac{(k-1)(\ell-1) + k + \ell - 2}{2} \equiv \frac{k-1}{2} + \frac{\ell-1}{2} \quad (\text{mod } 2).$$

Plugging this into (A.3.16) yields the inductive assertion. Thus the theorem is proved.   □

Proof of Proposition 6.3.2. We show that for $n \geq 3$ an odd integer we have

$$\left(\frac{2}{n}\right) = (-1)^{\frac{n^2-1}{8}}.$$

Again, we consider the prime factorization $n = p_1 \cdots p_r$ and prove the claim by induction on r. If $r = 1$, i.e., n is a prime number, the assertion is just Corollary A.3.2. For the induction step, assume $r \geq 2$, and that the claim is true for n' with fewer than r prime factors. Write $n = k\ell$ for numbers k, $\ell \geq 3$. By multiplicativity and the induction hypothesis, we have

$$\left(\frac{2}{n}\right) = \left(\frac{2}{k}\right) \cdot \left(\frac{2}{\ell}\right) = (-1)^{\frac{k^2-1}{8} + \frac{\ell^2-1}{8}}. \qquad (A.3.17)$$

Now $\frac{(k^2-1)(\ell^2-1)}{8}$ is divisible by 8, hence

$$\frac{n^2-1}{8} = \frac{(k^2-1)(\ell^2-1) + k^2 + \ell^2 - 2}{8} \equiv \frac{k^2-1}{8} + \frac{\ell^2-1}{8} \quad (\text{mod } 2).$$

$$(A.3.18)$$

If we plug this into (A.3.17), we obtain the inductive assertion. Thus the proposition is proved.   □

References

1. Adleman, L.M., and Huang, M.-D.A. *Primality testing and abelian varieties over finite fields*, Lecture Notes in Mathematics, Vol. 1512. Springer-Verlag, Berlin, Heidelberg, New York 1992.
2. Adleman, L.M., Pomerance, C., and Rumely, R.S., On distinguishing prime numbers from composite numbers. Ann. Math. 117 (1983) 173–206.
3. Agrawal, M., Kayal, N., and Saxena, N., PRIMES is in P. Preprint, http://www.cse.iitk.ac.in/news/primality.ps, August 8, 2002.
4. Agrawal, M., Kayal, N., and Saxena, N., PRIMES is in P. Preprint (revised), http://www.cse.iitk.ac.in/news/primality_v3.ps, March 1, 2003.
5. Alford, W., Granville, A., and Pomerance, C., There are infinitely many Carmichael numbers. Ann. Math. 139 (1994) 703–722.
6. Apostol, T., *Introduction to Analytic Number Theory*. 3rd printing. Springer-Verlag, Berlin, Heidelberg, New York 1986.
7. Artjuhov, M., Certain criteria for the primality of numbers connected with the Little Fermat Theorem (in Russian). Acta Arith. 12 (1966/67) 355–364.
8. Bach, E., Explicit bounds for primality testing and related problems. Inform. and Comput. 90 (1990) 355–380.
9. Baker, R.C., and Harman, G., The Brun-Titchmarsh Theorem on average. In: Berndt, B.C., *et al.*, Eds., *Analytic Number Theory, Proceedings of a Conference in Honor of Heini Halberstam*, pages 39–103. Birkhäuser, Boston 1996.
10. Bernstein, D.G., Proving primality after Agrawal, Kayal, and Saxena. Preprint, http://cr.yp.to/papers#aks, January 25, 2003.
11. Bernstein, D.G., Distinguishing prime numbers from composite numbers: The state of the art in 2004. Preprint, http://cr.yp.to/primetests.html#prime2004, February 12, 2004.
12. Bosma, W., and van der Hulst, M.-P., *Primality Proving with Cyclotomy*. PhD thesis, University of Amsterdam, 1990.
13. Bornemann, F., PRIMES is in P: A breakthrough for "everyman". Notices of the AMS 50 (2003) 545–552.
14. Burthe, R., Further investigations with the strong probable prime test. Math. Comp. 65 (1996) 373–381.
15. Cormen, T.H., Leiserson, C.E., Rivest, R.L., and Stein, C., *Introduction to Algorithms*. 2nd Ed. MIT Press, Cambridge 2001.
16. Crandall, R., and Pomerance, C., *Prime Numbers: A Computational Perspective*. Springer-Verlag, Berlin, Heidelberg, New York 2001.
17. Damgård, I., Landrock, P., and Pomerance, C., Average case error estimates for the strong probable prime test. Math. Comp. 61 (1995) 513–543.
18. Feller, W., *An Introduction to Probability Theory and Its Applications*, Vol. 1. 3rd Ed. Wiley, New York 1968.
19. Fouvry, E., Théorème de Brun-Titchmarsh; application au théorème de Fermat. Invent. Math. 79 (1985) 383–407.

20. Gauss, C.F., Disquisitiones Arithmeticae, 1801.
21. Graham, R.L., Knuth, D.E., and Patashnik, O., *Concrete Mathematics*. 2nd Ed. Addison-Wesley, Boston 1994.
22. Hardy, G., and Wright, E., *An Introduction to the Theory of Numbers*, 5th Ed. Clarendon Press, Oxford 1979.
23. Hopcroft, J.E., Motwani, R., and Ullman, J.D., *Introduction to Automata Theory, Languages, and Computation*. 2nd Ed. Addison-Wesley, Boston 2001.
24. Koblitz, N., *A Course in Number Theory and Cryptography*. 2nd Ed. Springer-Verlag, Berlin, Heidelberg, New York 1994.
25. Knuth, D.E., *Seminumerical Algorithms*. Volume 2 of *The Art of Computer Programming*. 3rd Ed. Addison-Wesley, Boston 1998.
26. Lehmann, D.J., On primality tests. SIAM J. Comput. 11 (1982) 374–375.
27. Lenstra, H.W., Primality testing with cyclotomic rings. Unpublished. August 2002.
28. Lidl, R., and Niederreiter, H., *Introduction to Finite Fields and Cryptography*. Cambridge University Press, Cambridge 1986.
29. Miller, G.M., Riemann's hypothesis and tests for primality. J. Comput. Syst. Sci. 13 (1976) 300–317.
30. Nair, M., On Chebyshev-type inequalities for primes. Amer. Math. Monthly 89 (1982) 126–129.
31. Niven, I., Zuckerman, H.S., and Montgomery, H.L., *An Introduction to the Theory of Numbers*. 5th Ed. Wiley, New York 1991.
32. Pinch, R.G.E., The Carmichael numbers up to 10^{15}. Math. Comp. 61 (1993) 381–391. Website: http://www.chalcedon.demon.co.uk/rgep/car3-18.gz.
33. Pinch, R.G.E., The pseudoprimes up to 10^{13}. In: Bosma, W., University of Nijmegen, The Netherlands (Ed.), Algorithmic Number Theory, 4th International Symposium, Proceedings, pages 459–474. Lecture Notes in Computer Science, Vol. 1838. Springer-Verlag, Berlin, Heidelberg, New York 2000.
34. Pratt, V., Every prime has a succinct certificate. SIAM J. Comput. 4 (1975) 214–220.
35. Rabin, M.O., Probabilistic algorithm for testing primality. J. Number Theory 12 (1980) 128–138.
36. Rivest, R., Shamir, A., and Adleman, L., A method for obtaining digital signatures and public-key cryptosystems. Comm. Assoc. Comput. Mach. 21 (1978) 120–126.
37. Salomaa, A., *Public-Key Cryptography*. 2nd Ed. Springer-Verlag, Berlin, Heidelberg, New York 1996.
38. Schönhage, A., and Strassen, V., Schnelle Multiplikation großer Zahlen. Computing 7 (1971) 281–292.
39. Solovay, R., and Strassen, V., A fast Monte-Carlo test for primality. SIAM J. Comput. 6 (1977) 74–86.
40. Stinson, D.R., *Cryptography: Theory and Practice*. CRC Press, Boca Raton 2002.
41. Von zur Gathen, J., and Gerhard, J., *Modern Computer Algebra*. 2nd Ed. Cambridge University Press, Cambridge 2003.

Index